Alexander Lotz

Designing an antimicrobial and cell-adhesive multilayer

Alexander Lotz

Designing an antimicrobial and cell-adhesive multilayer

How to counteract nosocomial infections associated with medical devices in hospitals

Südwestdeutscher Verlag für Hochschulschriften

Impressum/Imprint (nur für Deutschland/only for Germany)
Bibliografische Information der Deutschen Nationalbibliothek: Die Deutsche Nationalbibliothek verzeichnet diese Publikation in der Deutschen Nationalbibliografie; detaillierte bibliografische Daten sind im Internet über http://dnb.d-nb.de abrufbar.
Alle in diesem Buch genannten Marken und Produktnamen unterliegen warenzeichen-, marken- oder patentrechtlichem Schutz bzw. sind Warenzeichen oder eingetragene Warenzeichen der jeweiligen Inhaber. Die Wiedergabe von Marken, Produktnamen, Gebrauchsnamen, Handelsnamen, Warenbezeichnungen u.s.w. in diesem Werk berechtigt auch ohne besondere Kennzeichnung nicht zu der Annahme, dass solche Namen im Sinne der Warenzeichen- und Markenschutzgesetzgebung als frei zu betrachten wären und daher von jedermann benutzt werden dürften.

Coverbild: www.ingimage.com

Verlag: Südwestdeutscher Verlag für Hochschulschriften GmbH & Co. KG
Heinrich-Böcking-Str. 6-8, 66121 Saarbrücken, Deutschland
Telefon +49 681 37 20 271-1, Telefax +49 681 37 20 271-0
Email: info@svh-verlag.de

Approved by: Mainz, Johannes-Gutenberg Universität, Diss., 2012

Herstellung in Deutschland (siehe letzte Seite)
ISBN: 978-3-8381-3343-0

Imprint (only for USA, GB)
Bibliographic information published by the Deutsche Nationalbibliothek: The Deutsche Nationalbibliothek lists this publication in the Deutsche Nationalbibliografie; detailed bibliographic data are available in the Internet at http://dnb.d-nb.de.
Any brand names and product names mentioned in this book are subject to trademark, brand or patent protection and are trademarks or registered trademarks of their respective holders. The use of brand names, product names, common names, trade names, product descriptions etc. even without a particular marking in this works is in no way to be construed to mean that such names may be regarded as unrestricted in respect of trademark and brand protection legislation and could thus be used by anyone.

Cover image: www.ingimage.com

Publisher: Südwestdeutscher Verlag für Hochschulschriften GmbH & Co. KG
Heinrich-Böcking-Str. 6-8, 66121 Saarbrücken, Germany
Phone +49 681 37 20 271-1, Fax +49 681 37 20 271-0
Email: info@svh-verlag.de

Printed in the U.S.A.
Printed in the U.K. by (see last page)
ISBN: 978-3-8381-3343-0

Copyright © 2012 by the author and Südwestdeutscher Verlag für Hochschulschriften GmbH & Co. KG and licensors
All rights reserved. Saarbrücken 2012

"If I were to suggest that between the Earth and Mars there is a china teapot revolving about the sun in an elliptical orbit, nobody would be able to disprove my assertion provided I were careful to add that the teapot is too small to be revealed even by our most powerful telescopes. But if I were to go on to say that, since my assertion cannot be disproved, it is an intolerable presumption on the part of human reason to doubt it, I should rightly be thought to be talking nonsense. If, however, the existence of such a teapot were affirmed in ancient books, taught as the sacred truth every Sunday, and instilled into the minds of children at school, hesitation to believe in its existence would become a mark of eccentricity and entitle the doubter to the attentions of the psychiatrist in an enlightened age or of the Inquisitor in an earlier time."

– *Bertrand Russell*

…for Milena

Abstract

In this work surface modifications were developed that bear cell-adhesive as well as antimicrobial properties. Fast cell adhesion and wound healing is desired for biomaterials since otherwise the material would be recognized as a foreign body and infectious bacteria could enter the cavity between material and tissue. Plasma polymerization was used as the deposition technique since a broad range of materials can be coated independent of their composition. As a cell-adhesive coating plasma polymerized allylamine was chosen since it is cell-friendly and offers a platform for wet chemical modifications such as grafting of fibronectin. Furthermore, it serves as a barrier layer for zinc and silver containing sub-layers that exhibit antimicrobial properties due to the release of zinc and silver. The layer systems were spectroscopically and microscopically investigated. Cell-adhesive and antimicrobial properties were tested with various types of cells and bacteria.

Acknowledgments

I want to thank Dr. Renate Förch who supervised me at the Max Planck Institute for Polymer Research and gave me the chance to take responsibility in various projects. Thanks also to Peter Cierniak from the University of Cologne who introduced me to bacterial testing and supported my work during my visits to Cologne. I also would like to thank Nina Dohm from the University Medical School in Mainz for her help on testing the longevity of antimicrobial properties. For his help on working with endothelial cells I would like to thank Martin Heller from the MPI-P in Mainz. Also, I want to thank Dr. Juan Carlos Ruiz and the École Polytechnique Montréal for their help with XPS. Kerstin Malzahn and Mark Bedrich I would like to thank for proof reading the manuscript.

A big thank you to my wife who endures me when I am grumpy…

Table of Contents

Abstract 2

Acknowledgments 3

Table of Contents 5

Table of Figures 7

List of Tables 10

Table of Abbreviations 11

1   Introduction and Aim of this Work 13
2   Materials and Methods 21
2.1   Plasma Enhanced Chemical Vapor Deposition 21
2.2   Physical Vapor Deposition 27
2.3   Step Profiling 28
2.4   Contact Angle Goniometry 29
2.5   Infrared Spectroscopy 30
2.5.1   Infrared Reflection Absorption Spectroscopy 31
2.5.2   Attenuated Total Reflection Infrared Spectroscopy 31
2.6   Immobilization of Biomolecules 32
2.7   Surface Plasmon Resonance 33
2.8   Inductively Coupled Plasma Optical Emission Spectroscopy 35
2.9   Scanning Electron Microscopy 37
2.10   Energy-Dispersive X-Ray Spectroscopy 37
2.11   X-Ray Photoelectron Spectroscopy 38
2.12   Overview over the Surface Analysis Techniques 40
2.13   Bacterial Assays 41
2.14   Cell Microscopy 44
3   Cell-Adhesive Coatings with ppAA 46
3.1   Analysis of Cell-Adhesive Coatings 46
3.1.1   Film Adhesion Improvement with ppHMDSO 47
3.1.2   Analysis of ppAA Films 49
3.1.3   Biofunctionalization of ppAA 54
3.1.4   Cell Adhesion on Biofunctionalized ppAA Films 59

4 Antimicrobial Coatings with ppZn(acac)$_2$ 64
4.1 Analysis of ppZn(acac)$_2$ Films 64
4.1.1 Plasma Deposition of ppZn(acac)$_2$ 64
4.1.2 Zinc Content in ppZn(acac)$_2$ Films 68
4.2 Antimicrobial Efficacy of ppZn(acac)$_2$ Films 68
4.2.1 Optical Density of Bacterial Suspensions 69
4.2.2 Coating Woven Fabrics with ppZn(acac)$_2$ 70

5 Bilayered Coatings with ppAA and ppZn(acac)$_2$ 72
5.1 Plasma Deposition of ppAA on ppZn(acac)$_2$ 72
5.2 Release from Bilayered Coatings with ppAA on ppZn(acac)$_2$ 76
5.3 Antimicrobial Efficacy of Bilayers with ppAA on ppZn(acac)$_2$ 77
5.3.1 Bacterial Colony Counting from Bilayers with ppAA on ppZn(acac)$_2$ 78
5.3.2 Coating Urethral Catheters with Bilayers of ppAA on ppZn(acac)$_2$ 79
5.3.3 Cell Adhesion on Bilayers with ppAA on ppZn(acac)$_2$ 80
5.3.4 Cell Adhesion on Biofunctionalized Bilayers with ppAA on ppZn(acac)$_2$ 83

6 Multilayered Coatings with silver, ppZn(acac)$_2$, and ppAA 86
6.1 Film Analysis of Multilayered Coatings with silver, ppZn(acac)$_2$, and ppAA 87
6.2 Release of Silver and Zinc from Multilayered Coatings with silver, ppZn(acac)$_2$, and ppAA 92
6.3 Antimicrobial Efficacy of Multilayered Coatings with silver, ppZn(acac)$_2$, and ppAA 95
6.3.1 Bacterial Testing for *E. coli* on Multilayered Coatings with silver, ppZn(acac)$_2$, and ppAA 95
6.3.2 Coating Non-Woven Fabrics with Multilayered Coatings with silver, ppZn(acac)$_2$, and ppAA 96
6.3.3 Longevity of Antimicrobial Efficacy of Multilayered Coatings with silver, ppZn(acac)$_2$, and ppAA 98

7 Conclusion and Outlook 102

References 106

Table of Figures

Figure 1-1. Biofilm formation and proceeding stages. Planktonic bacteria can adsorb to a surface. After propagation they can form micro-colonies and secrete a polysaccharide matrix. This structure is called a biofilm. 15

Figure 1-2. Schematic representation of the multilayered system made from a thermally evaporated silver film, ppZn(acac)$_2$, ppAA as a barrier layer, and biomolecules immobilized on the barrier layer. 19

Figure 2-1. Various processes involved in a plasma deposition 23

Figure 2-2. Schematic representation of a plasma polymer network depicting hydrophobic ppHMDSO as an example. 24

Figure 2-3. Scheme of the plasma chamber used throughout this work. 26

Figure 2-4. Chemical structures of the monomers used throughout this work. 27

Figure 2-5. Scheme of the formation of a self-assembled monolayer from allyl mercaptane and subsequent activation by plasma (asterix). 28

Figure 2-6. Basic mechanism of step profiling. The stylus can move freely in the z-direction and its position is converted into depth information. 29

Figure 2-7. Scheme of a static sessile drop water contact angle measurement. 30

Figure 2-8. Schematic representation of the mode of function in IRRAS. 31

Figure 2-9. Scheme of the deposition and biofunctionalization of ppAA. 33

Figure 2-10. Basic scheme of the working principle of SPR measurements. 34

Figure 2-11. Example of two plasmon resonances recorded. The signal on the right (black line) shows a higher resonance angle resulting from a change in thickness and refractive index of an adsorbed layer. 35

Figure 2-12. Basic principle of the light detection from a plasma torch fed with an aerosol of a liquid sample. The light emitted can be analyzed in a detector. 36

Figure 2-13. Principle of XPS with electron transition. In the bound state the electron absorbs energy and bears a specific kinetic energy as a free electron. 39

Figure 2-14. Derivatization reaction of primary amines from a ppAA plasma deposit (R) with 4-(trifluoromethyl) benzaldehyde. 40

Figure 3-1. Scheme of the layer system discussed in chapter 3. 46

Figure 3-2. Film thickness distribution throughout the reactor after 5 s of HMDSO deposition. The precursor inlet is situated on the left whereas the vacuum pump is placed on the right. 47

Figure 3-3. IRRAS spectra of ppHMDSO with (red) and without (green) subsequent plasma activation with oxygen. Water contact angles are shown by the side view images of a water droplet on the surfaces (also refer to Table 3-2 in section 3.1.2). 48

Figure 3-4. Film thickness distribution throughout the reactor after 1 min of allylamine deposition. The monomer is located on the left whereas the pump is placed on the right. ... 49

Figure 3-5. IRRAS spectra of three different positions in the reactor and three different input powers (y-axis). Spectra were normalized to the strongest signal at 1650 cm^{-1}. The signals marked in red are representative for aliphatic hydrocarbon structures and are discussed in the context of loss of amino functionality. All spectra were background corrected manually to ensure comparability. 50

Table of Figures

Figure 3-6. Normalized IRRAS spectra of ppAA on ppHMDSO. Colored arrows mark signals of the respective films (details are shown in Figure 3-3 and Table 3-1). The according water contact angles are shown next to the spectra (also refer to Table 3-2 in section 3.1.3). 53

Figure 3-7. ATR-FTIR spectra of ppAA on PTFE. Also shown are spectra of PTFE and ppAA only. Distinct signals of ppAA are highlighted (see Table 3-1 for details). The according water contact angles are shown next to the spectra (also refer to Table 3-2 in this section). 54

Figure 3-8. Kinetic SPR measurements of fibronectin adsorption to ppAA a) with and b) without DEGDGE. The angular position (corresponding to adsorption or desorption processes) is plotted against the time. 56

Figure 3-9. IRRAS spectra of the wet chemically modified plasma deposit. The green box marks the detailed section shown in Figure 3-10. 57

Figure 3-10. Detailed section from Figure 3-9. The distinct signals are highlighted with red lines. 58

Figure 3-11. Calcein stained HUVECs after 24 h and 3 d on PTFE samples with various surface modifications based on ppAA deposits. 61

Figure 3-12. Cell counts evaluated from pictures shown in Figure 3-11. 62

Figure 3-13. Cell coverage derived from pictures shown in Figure 3-11. 63

Figure 4-1. Temperatures in the reactor (at the inner reactor wall surface and at the position where the samples were placed) versus the setpoint of the temperature controller. The straight line represents the angle bisector between x- and y-axis (a theoretical perfect heat convection). 65

Figure 4-2. Film thickness distribution throughout the reactor after 30 min of $ppZn(acac)_2$ deposition. 65

Figure 4-3. Film thickness of $ppZn(acac)_2$ within a petri dish of 3.5 mm diameter going from the rim to the center. 67

Figure 4-4. Optical density of bacterial suspensions of *S. aureus* and *P. aeruginosa* after incubation in uncoated and coated polystyrene petri dishes overnight. 69

Figure 4-5. Photographic images of a) uncoated and b) $ppZn(acac)_2$ coated woven PET from Sefar. 70

Figure 4-6. EDX spectra of the woven PET samples shown in Figure 4-5. 71

Figure 4-7. Live/Dead staining of *S. aureus* on a) uncoated and b) $ppZn(acac)_2$ coated woven PET shown in Figure 4-5. 71

Figure 5-1. Scheme of the layer system discussed in chapter 5. 72

Figure 5-2. Film thickness of ppAA inside a petri dish of 3.5 cm in diameter (from the rim to the center). 73

Figure 5-3. IRRAS spectra of ppAA films of varying thickness on $ppZn(acac)_2$. Distinct differences in spectra are highlighted by dashed lines. 75

Figure 5-4. Concentration of zinc in the leachate from bilayers of ppAA on $ppZn(acac)_2$ over the course of 120 h. 77

Figure 5-5. Plating of bacterial suspensions of *E. coli* after incubation overnight on a) uncoated and coated glass slides; b) $ppZn(acac)_2$ and c) a bilayer of ppAA on $ppZn(acac)_2$. A serial dilution was made from 20 µl with 1/1 (top), 1/2500, 1/5000, and 1/10000 (left to right). 78

Figure 5-6. Colony counting of *S. aureus* after incubation overnight on uncoated and coated polystyrene petri dishes ($ppZn(acac)_2$ and a bilayer of ppAA on $ppZn(acac)_2$). 79

Table of Figures

Figure 5-7. Colony counting of *S. aureus* incubated in a falcon tube together with uncoated and coated silicone catheter samples. 80

Figure 5-8. HUVECs on uncoated and coated titanium samples after 24 h. 81

Figure 5-9. Cell count evaluated from images shown in Figure 5-8. 82

Figure 5-10. Cell coverage evaluated from pictures shown in Figure 5-8. 83

Figure 5-11. *Fibroblasts NIH 3T3* on uncoated and coated glass cover slips after 24 hours of incubation. 84

Figure 5-12. Cell count evaluated from images shown in Figure 5-11. 85

Figure 6-1. Scheme of the layer system discussed in chapter 6. 86

Figure 6-2. Silver content in the supernatant on thermally evaporated silver with varying thickness. 87

Figure 6-3. SEM cross section of a Si-wafer coated with 50 nm of silver (black arrow), 500 nm of ppZn(acac)$_2$, and 20 nm ppAA. The barrier layer of ppAA was not resolved possibly due to lack of contrast. 88

Figure 6-4. EDX cross section of the sample shown in Figure 6-3. Nitrogen is not shown because of the amount of background signal observed for nitrogen. 89

Figure 6-5. XPS survey spectra of the single layer systems showing elemental composition (normalized to 100%) in atomic percent. Error bars for each element extend into the corresponding column for that element. 90

Figure 6-6. Zinc content in the leachate of a multilayer consisting of 50 nm of silver, ppZn(acac)$_2$ of varying thickness, and 20 nm ppAA. 94

Figure 6-7. Silver content in the leachate of a multilayer consisting of 50 nm of silver, ppZn(acac)$_2$ of varying thickness, and 20 nm ppAA. 94

Figure 6-8. Agar plates after incubation with supernatant containing *E. coli*. The supernatant was previously incubated on a) uncoated petri dishes, b) uncoated glass slides, c) Ag/ppZn(acac)$_2$, and d) Ag/ppZn(acac)$_2$/ppAA/FN coated glass slides. The two rows correspond to two dilutions of the supernatant. 96

Figure 6-9. Photographs of a) uncoated and b) multilayer coated non-woven PET. The multilayer consisted of silver + ppZn(acac)$_2$ + ppAA + linker bound fibronectin. 97

Figure 6-10. EDX spectra of the woven PET samples shown in Figure 6-9. 97

Figure 6-11. Live/Dead staining of *S. aureus* on a) uncoated and b) coated woven PET shown in Figure 6-9. 98

Figure 6-12. Qualitative evaluation of the longevity of the antimicrobial efficacy for various coatings on PTFE on *S. aureus*. 99

Figure 6-13. Qualitative evaluation of the longevity of the antimicrobial efficacy for various coatings on PTFE on *P. aeruginosa*. 101

List of Tables

Table 2-1. Monomer flow rates, input powers, and deposition times. 26
Table 2-2. Overview over the surface analysis techniques used in this work. 41
Table 3-1. IR signals of ppAA 51
Table 3-2. Water contact angles of various surfaces used in the context of cell adhesion on ppAA. 59
Table 4-1. Water contact angles of various surfaces used in the context of antimicrobial coatings with ppZn(acac)$_2$. 67
Table 4-2. Zinc content per volume element of ppZn(acac)$_2$ films. 68
Table 5-1. Zinc content per volume element of ppZn(acac)$_2$ films modified with top-layers of ppAA in varying thickness 74
Table 5-2. Water contact angles of various surfaces used in the context of bilayered coatings from ppZn(acac)$_2$ and ppAA. 75
Table 6-1. The original data set for Figure 6-5 with elemental composition in percent without normalization (difference to 100% is attributed to impurities, e.g. Na, Cl, Si) 91

Table of Abbreviations

AA	allylamine
ATR	attenuated total reflection
CW	continuous wave
DC	duty cycle
DMEM	Dulbecco's modified eagle medium
EDX	energy-dispersive X-ray spectroscopy
HMDSO	hexamethyldisiloxane
ICP-OES	inductively coupled plasma optical emission spectroscopy
IR	infrared
IRRAS	infrared reflection absorption spectroscopy
OD_{600}	optical density measured at $\lambda = 600$ nm
pp	plasma polymerized
PECVD	plasma enhanced chemical vapor deposition
PEG	polyethylene glycol
PET	polyethylene terephthalate
PTFE	polytetrafluoroethylene
PVD	physical vapor deposition
SAM	self-assembled monolayer
SEM	scanning electron microscopy
SPR	surface plasmon resonance
XPS	X-ray photoelectron spectroscopy
$Zn(acac)_2$	zinc acetylacetonate

1 Introduction and Aim of this Work

Modern medicine utilizes many different artificial materials to compensate for defects that can occur in a human body such as fractures, burned skin, open wounds, loss of teeth, et cetera. In order to provide a substitute for tissue or to support damaged tissue artificial materials are inserted into the body and are generally termed biomaterials.[1] Different aspects need to be taken into account; among them chemical and mechanical stability, interaction with the surrounding tissue, or immune response to name but a few.[2] Bacterial infections emerge frequently in medical care with regard to biomaterials or medical devices, e.g. burn wounds, dental implants, or urethral catheters,[3-5] and bacterial colonization is the main factor for implant failure.[6]

Microorganisms like bacteria can be found in large quantities in nature and even within and on the human body.[7] While most of the bacteria are non-pathogenic some types or strains do develop harmful properties and cause millions of fatalities throughout the world.[8] This problem is not restricted to developing countries where lack of hygiene may play an important role. Especially harmful and pathogenic bacteria with resistance to many antibiotics can be found in hospitals worldwide.[9] Hospital acquired infections are very common due to the high number of patients and the overuse and preventive application of antibiotics that cause development of bacterial resistance.[10,11] Infections are among the most common causes of deaths with bacteria being the main contribution.[12] Two of the most prominent types of bacteria that are causing hospital acquired infections are *Staphylococcus aureus (S. aureus)*[13] and *Pseudomonas aeruginosa (P. aeruginosa)*[14].

S. aureus is a gram-positive bacteria found on human skin as well as on nasal soft tissue.[15] It is able to prevent immunological strategies by secretion of staphyloxantin. This carotinoid acts as a scavenger for reactive oxygen species which the host's immune system produces to prevent infection.[16] Certain strains of *S. aureus* like *MRSA* (methicillin resistant *S. aureus*) are found in hospitals worldwide and show resistance against many different antibiotics. It can grow under anaerobic conditions which enables it to colonize wounds and biomaterials embedded in tissue and even survives under dry conditions.[17,18] Virulence factors of *S. aureus* are diverse and often associated with skin related diseases but may also include toxic shock syndrome with fatal outcome within hours if left untreated.[19]

P. aeruginosa is a gram-negative bacterium found in water and soil and can as well grow under anaerobic conditions. Its robustness is increased by formation of biofilms.[20,21] Like many gram-negative bacteria *P. aeruginosa* exerts resistance to several antimicrobials and antibiotics.[22] It is found in hospitals even in distilled water or disinfectants and is also responsible for nosocomial infections.[23,24] Whereas the exterior of gram-positive bacteria consists of a plasma membrane and a peptidoglycan layer, gram-negative bacteria have an additional outer membrane. This outer membrane plays an important role in resistance since it provides a further barrier that helps to prevent incorporation of antimicrobial substances into the bacterium either passively or actively via transport mechanisms.[25] It is also the reason why such bacteria survive contact with quaternary ammonium compounds that are generally known to be antimicrobial by incorporation into membrane structures.[26]

Solutions to hospital acquired infections with these types of bacteria need to be sought, not only because of the high morbidity of patients. Also, high costs for the health care system due to intensive treatment and prolonged stay in hospital, possibly with intensive care, are an economical factor.[27] Transfer of bacteria most likely happens on surfaces and interfaces[28] and can include medical devices, tools, food, clothing, and even the skin of personnel as well as other patients.[29-31] This work will address various materials and devices used in medical fields such as PTFE (polytetrafluoroethylene, used for vascular grafts), fabrics from PET (polyethylene terephthalate, used as wound dressing material), titanium (used for osteosynthesis plates in trauma surgery or dental implants), and silicone (used for urethral catheters).

Often pathogenicity of bacteria colonizing a surface involves the formation of a biofilm.[32] A biofilm is a community of bacteria which secrete biomolecules to form a biomacromolecular structure. Bacteria can drastically change their behavior in such a biofilm, produce various toxins and become increasingly unaffected by many different antimicrobial agents.[33-36] This problem is especially prominent in cases where such surfaces come in contact with tissue since the high flow of bodily fluids and nutrients accelerate biofilm formation. As illustrated in Figure 1-1, the formation of a biofilm involves various steps beginning with the reversible attachment of single bacteria in solution (planktonic state) to a surface, irreversible attachment by secretion of polysaccharides, micro-colonization and propagation, and finally the fully matured biofilm.[37,38] After the initial passive and reversible attraction to surfaces bacteria actively use saccharide adhesins for surface attachment.[39-41] Response towards antimicrobial agents is also faster in a biofilm compared to planktonic bacteria. This is thought to be due to intercellular communication, so-called quorum sensing, which is enhanced by the proximity of the bacterial cells in a biofilm.[42,43] Also, genetic information is easily exchanged within a colony of bacteria and passed down to the whole community to increase resistance towards antimicrobial agents.[44]

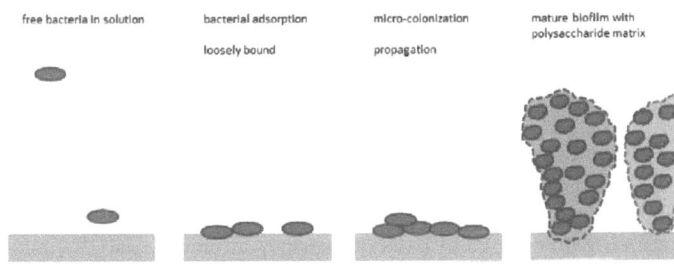

Figure 1-1. Biofilm formation and proceeding stages. Planktonic bacteria can adsorb to a surface. After propagation they can form micro-colonies and secrete a polysaccharide matrix. This structure is called a biofilm.

If an infection is to be inhibited at all, the highest chance of doing so will be in the initial phase where the bacterium is in its most vulnerable state.[45] Hence, surfaces that bear antimicrobial properties would be most promising to prevent bacterial colonization and therefore favored over systemic approaches on the patient.

Antimicrobial properties for surface coatings are of high interest in the health care system.[9] Different approaches are possible to create such coatings including synthesis of antifouling surfaces,[46] antimicrobial peptides,[47] quaternary ammonium compounds,[48] bacteriophages,[49] or metal release systems[50] to name but a few. Antifouling surface modifications can be made with various chemical approaches like polymer brushes, PEG-like structures, superhydrophilic, and superhydrophobic modifications as well as lotus effect approaches. Results are somewhat contradictory concerning the effect of wettability.[46,51-55] The mode of action is yet unclear and passive anti-fouling surfaces can eventually be colonized due to secretion of extracellular material either by the host tissue of the patient or by bacteria forming a biofilm on the medical device.[56]

Quaternary ammonium compounds, antimicrobial peptides, and immobilized bacteriophages are promising due to their known mode of action. Nevertheless, they can only act on bacteria in contact with the surface. They are prone to coverage by extracellular material and may not exert their properties for long in living tissue. Quaternary ammonium compounds are thought to act on ion channels in the bacterial membrane and can interact with the lipid bilayer involving their alkyl chains, thus causing loss of structural integrity. However, resistance in some bacteria is known.[57,58] Bacteriophages have the drawback that they are very host specific and a whole cocktail of various bacteriophages needs to be immobilized on a surface in order to span a whole range of bacterial species targeted. Also, the patient's immune system is able to recognize the phages and produces antibodies rendering bacteriophage treatment 'a last resort' because it can only be used once for a given patient.[59]

Release systems can be tailored by incorporation of antimicrobial compounds into the surface such as pH- or temperature-sensitive hydrogels or other swollen polymeric networks and fabrics.[60,61] Antimicrobial substances can leave the coating and exert their properties in solution. This approach is not depending on contact killing and may very well include triggered active release, e.g. of antibiotics.[62]

In the same manner metal release systems are not relying on contact killing either but actively release material into the surrounding medium. Such surfaces could increase the concentration of metal up to a bactericidal level. The most commonly used metal for antimicrobial devices is silver due to its well-known bactericidal effects and low number of resistant bacteria.[63,64] The complete mode of action of metals in bacteria is yet unknown but involves oxidative stress and change of protein structures.[65-68] The latter is possible because silver reacts with sulfur present in proteins. It can thus deactivate enzymes by structural degeneration and denaturation.[69,70] Silver as well as zinc can also prevent readout of genetic information.[71,72] Zinc furthermore reduces catabolic reactions like the synthesis of peptides.[73,74] Many metals not only exhibit antimicrobial but also cytotoxic effects, e.g. silver, copper, and zinc.[75-77] Zinc is a promising candidate for an antimicrobial metal releasing system since it is essential to humans and does show lower cytotoxicity compared to silver.[78,79] Also, combinations of release systems and contact killing surfaces would present a more effective route due to the diversity of the multiple actions and their synergistic effect on the bacteria.[80]

Preparation of surface modifications with metal containing films include dip coating, grafting of complexes, magnetron sputtering, and physical vapor deposition.[81-85] A controlled release of metal into the surroundings of the coated device would be favorable and could also help in prolonging the effect of the coating since the reservoir of metal is limited. Incorporation of metal particles into polymeric material is a possible pathway to control the release.[86] Particles in deeper layers of the surface show a slower release of material into the surrounding medium. Another approach is the deposition of plasma polymers and subsequent immersion in a metal containing solution. The films are soaked with metal ions and can then be exposed to a reducing agent, thus causing the formation of nanoparticles. Metal is then released upon contact with body fluids.[87] Plasma polymers seem to provide a good basis for such an approach and for release systems in general since they can swell by absorption of water. At the same time they exhibit high crosslinking and therefore stability towards solvents.[88,89] Barrier layers from plasma polymers are also known to influence the permeability of membranes for gases depending on the degree of crosslinking and thickness.[90-93] Metal containing plasma polymers from metal-organic precursors like zinc acetylacetonate ($Zn(acac)_2$) are used for microelectronics.[94] But they are also known to release their metal content over time upon immersion in aqueous solutions. Such plasma polymers could exhibit antimicrobial properties but the release is not controlled and properties are lost within a few hours.[95]

The aim of this work is to modify such metal containing surfaces with a top-layer made from another plasma polymer so that the release of metal is reduced and therefore prolonged. A plasma polymer film made from allylamine is suitable since those films are known to exhibit cell-friendly properties.[96] The layer system presented in this work consists of an antimicrobial layer made from plasma polymerized zinc acetylacetonate ($ppZn(acac)_2$) and a top-layer of plasma polymerized allylamine (ppAA). Such a bilayer can further be modified by immobilization of biomolecules on ppAA which is possible because of the amount of amino functional moieties that ppAA is providing.[97,98]

Up to this point the complete bilayer consists of plasma polymers; a zinc-containing metal-organic layer and a cell-adhesive top-layer made from ppAA. To broaden the spectrum of antimicrobial agents another layer underneath the bilayer system can be added as presented in Figure 1-2. In this case a thermally evaporated silver film is added in between the substrate and the bilayer. This is thought to minimize development of resistance in bacteria due to the multiple bactericidal approaches. It also offers a route to address bacteria resistant to zinc. The fact that the silver layer is placed beneath the bilayer can help to prevent cytotoxic effects due to retardation of the silver release.

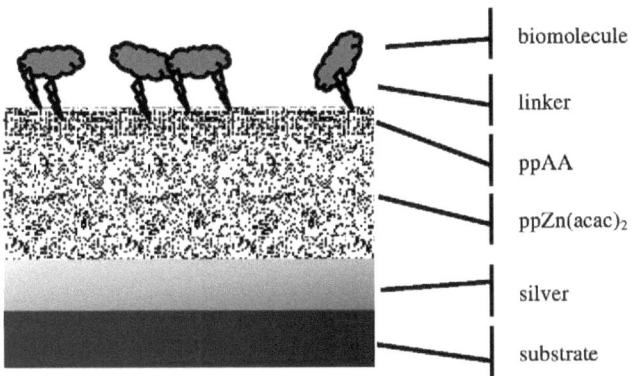

Figure 1-2. Schematic representation of the multilayered system made from a thermally evaporated silver film, ppZn(acac)2, ppAA as a barrier layer, and biomolecules immobilized on the barrier layer.

Cell adhesion is unspecifically promoted by ppAA[99] but coupling biomolecules to ppAA offers a pathway to specifically target biological systems.[97] Considering the fact that foreign body reactions and poor wound healing seem to be the main factors for bacterial infections and subsequent implant failure, immobilization of biomolecules that promote cell adhesion to the surface should be considered.[100-102] A suitable approach is the utilization of proteins present in the extracellular matrix. The extracellular matrix is responsible for interconnection between cells and is the main cause of the structural integrity of tissue.[103] It contains protein structures like collagen, laminin, and elastin in a proteoglycane matrix. One key element for the adhesion of cells to the matrix is the interaction between membrane-bound integrins and fibronectin.[104] Fibronectin does not only support cell motility but does also play a vital role in wound healing.[105] Fibronectin is a dimeric protein with various binding sites. Among them is a loop structured domain containing the amino acid sequence

RGD (arginine-glycine-aspartic acid).[106,107] This sequence is thought to mainly contribute to integrin binding.

Immobilization of fibronectin mimics the natural environment of an extracellular matrix and quick attachment of various cell types could accelerate wound healing and thus integration of medical devices within the tissue.[108-110] Close proximity of a protein to the surface could lead to denaturation but it is known that denatured fibronectin still exhibits its properties and it was found that physisorption of the protein alone already increases cell adhesion.[107,111-113] Nevertheless, chemisorption would offer higher stability especially in cases where high shear forces play a role, e.g. in vascular grafts and stents. In such devices a high flow of liquid is present and is causing high mechanical stress.[114] Therefore, a binding chemistry is applied that involves the utilization of a short PEG-like linker molecule.[113,115]

Together with the antimicrobial properties of the aforementioned layers these cell-adhesive surface modifications could lead to fast wound healing and would reduce the risk of implant failure. This work will focus on the design and construction of these multilayers on various materials and medical products. Different combinations of the layers will be discussed and the release of metal as well as the behavior of mammalian cells and bacteria on coated and uncoated surfaces will be investigated.

2 Materials and Methods

This section will focus on the different methodologies utilized throughout this work. It is mostly in logical/chronological order of the procedures conducted and is held as concise as possible. Each method section includes a short description of the theoretical background of the technique, a summary of the materials used as well as the experimental procedures.

2.1 Plasma Enhanced Chemical Vapor Deposition

Plasma is often referred to as the 4th state of matter in which the atoms or molecules are at least partially separated into ions and free electrons and has been first described by Langmuir in 1928.[116] Plasma is by far no rare condition and in fact is the most common state of matter throughout the known universe.[117] Various descriptions for the plasma state can be given which are dependent on plasma pressure, temperature, and equilibrium. In contrast to the processes in our sun, plasma in the presented work refers to a low-pressure, non-equilibrium, cold plasma.[118] Working at low pressure provides improved control over the composition of the plasma. Non-equilibrium goes hand in hand with low temperature and means that the electron temperature by far exceeds that of the other species, namely excited neutrals, ions, and radicals, which stay at or close to room temperature.[118]

Plasma is easily created by feeding a gas into an evacuated chamber and applying an externally driven oscillating electric current through an electrode surrounding the chamber. This electrode is either a simple ring electrode or a grounded copper coil. The alternating current creates an alternating electromagnetic field inside the chamber. The electrons in the gas phase will respond to this external electromagnetic field and oscillate along with it. Depending on the frequency this oscillation is too fast for the core of the molecules and while they do not change their kinetic energy, that of the electrons will increase to a level enabling them to escape their energetically favored state as bound electrons. Nevertheless, free electrons can be scavenged by radicals and ions releasing energy by emission of light commonly referred to as glow discharge. All exited species can interact and undergo chemical reactions, thus forming species with growing molecular weight.[119] At the same time any solid material in contact with the plasma forms excited species on the surface as well and reactions with species in the plasma state are possible. Continuous deposition of such species leads to solid state networks and thin films in the nano- to micrometer range.[120] This process is commonly known as plasma polymerization or plasma enhanced chemical vapor deposition.

Taking some simple examples, general reactions can be derived:[118]

ionization by electron impact	$Ar + e^-$	$\rightarrow$	$Ar^+ + 2e^-$
excitation by electron impact	$Ne + e^-$	$\rightarrow$	$Ne^* + e^-$
de-excitation and ionization	$Ne^* + Ar$	$\rightarrow$	$Ne + Ar^+ + e^-$
ionization/dissociation of a molecule	$Cl_2 + e^-$	$\rightarrow$	$Cl^- + Cl$
electron capture and recombination	$Ar^+ + e^-$	$\rightarrow$	$Ar + h\nu$

Polymerizable species can contribute to growth mechanisms all of which are possible during one plasma polymerization process:[118]

$$M_n^+ + M \rightarrow M_{n+1}^+$$

$$M_n^- + M \rightarrow M_{n+1}^-$$

$$M_n^\cdot + M \rightarrow M_{n+1}^\cdot$$

$$M_n^* + M \rightarrow M_{n+1}^*$$

Further reactions and rearrangements are possible especially for organic molecules. One has to keep in mind that due to the large number of possible species and the high number of possible interactions as well as further ion bombardment the resulting deposit will bear complex film chemistry. This is in contrast to wet chemically synthesized polymers. Figure 2-1 gives an overview over the basic processes that are involved in a plasma deposition. These include excitation of the gas, formation of polymer forming intermediates, and reaction of these intermediates on the surface. The excited species can also cause creation of reactive species on the surface (responsible for the interaction of the substrate and the deposited film) but may also cause etching of the substrate as well as of the polymer film. Such surface activation processes are the main reason for good adhesion between most substrates and the plasma polymer film and are considered the main advantage over wet chemical approaches.

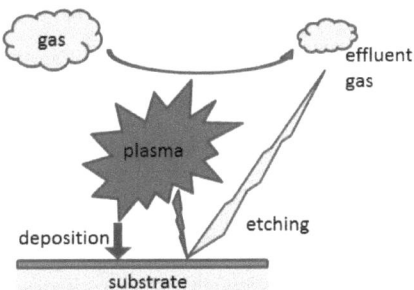

Figure 2-1. Various processes involved in a plasma deposition

With increasing number and complexity of the species in the gas and plasma phase the complexity of the film structure increases too. Plasma from noble gases most likely, but not exclusively, causes etching processes and does not take part in chemical reactions on the surface. However, noble gas plasma induces reactions via energy transfer upon impact with the surface. Depending on the substrate this can already cause a very diverse surface chemistry. The situation becomes more complicated for gasses that easily react as in the case of oxygen. Chemical pathways of organic precursors in plasma are therefore almost impossible to simulate or to predict a priori. Figure 2-2 shows a hypothetical composition of a plasma polymerized hexamethyldisiloxane film (ppHMDSO) as one of the most widely known plasma polymers.

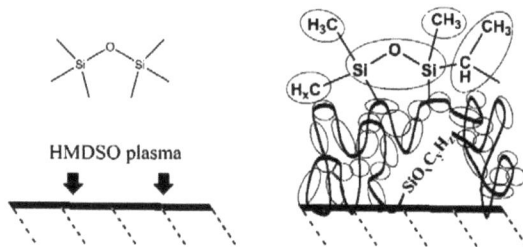

Figure 2-2. Schematic representation of a plasma polymer network depicting hydrophobic ppHMDSO as an example.

The composition of the resulting film is mainly determined by the power density which is strictly determining the energy available for the gas molecules. Higher input power as well as lower gas flow means higher energy density. Thus, more energy is available to break up bonds in a molecule and to open new pathways of interaction. Lower input power offers the advantage of structural retention of the precursor molecule such that functional groups may be available for further wet chemical reactions after plasma polymerization. Moreover, instead of applying the power continuously (continuous wave mode or CW) the plasma can be driven in duty cycles meaning that the plasma is turned on and off alternatingly in the micro- to millisecond range. During the plasma-on times (t_{on}) molecules absorb energy and undergo structural reformation during plasma-off times (t_{off}). In order to express the power input a duty cycle DC is defined as:

$$\text{DC} = \frac{t_{on}}{t_{on} + t_{off}} \qquad 1.1$$

$$P_{eq} = DC \cdot P_{in} \qquad 1.2$$

In this case the equivalent power P_{eq} is determined by the product of the power at which the electrode is driven P_{in} and the duty cycle DC[1].

The main advantage of plasma-induced deposition is the generation of reactive species on the surface and therefore good adhesion of the resulting film on substrates with diverse surface chemistry is achieved. Radicals on the activated surface that are readily reacting with species from the gas or plasma phase play a major role in film stability and adhesion.[121,122] Furthermore, a non-equilibrium plasma does not influence the bulk properties of the substrate and merely acts on the surface thus offering a non-destructive coating procedure.[118] It is a fast process (usually in the range of minutes) and film thickness is easily controlled by deposition time. Nevertheless, the complexity of chemical pathways renders plasma polymerization a procedure that is time consuming in terms of surface analysis. Also, the fact that it is a vacuum process is causing problems in upscaling the procedures.

Figure 2-3 shows a schematic representation of the basic setup used in this work. The chamber consists of a 30 cm long Pyrex glass tube with a diameter of about 10 cm. There are inlet valves through which various gases can be fed. The flow of standard gases like argon and oxygen can be controlled with mass flow controllers (MKS 1179B12CS1BV). Chamber pressure is adjusted with a Leybold D16 BCS vacuum pump and is gauged with an MKS Baratron manometer. Base pressure was determined and calibrated to 10^{-6} bar. The pump is protected from any effluent gasses with a cold trap cooled with liquid nitrogen. Plasma was driven by a 13.56 MHz function generator delivering up to 150 W of input power. The impedance can be adjusted with a matching box. The electrode is driven with shielded coaxial cables and is either an isolated copper coil wound around the chamber or a single ring electrode. The latter was used for all $Zn(acac)_2$ depositions.

---

[1] This expression is mathematically awkward since it might be taken for a term of two variables. Nevertheless, it is the common terminology.

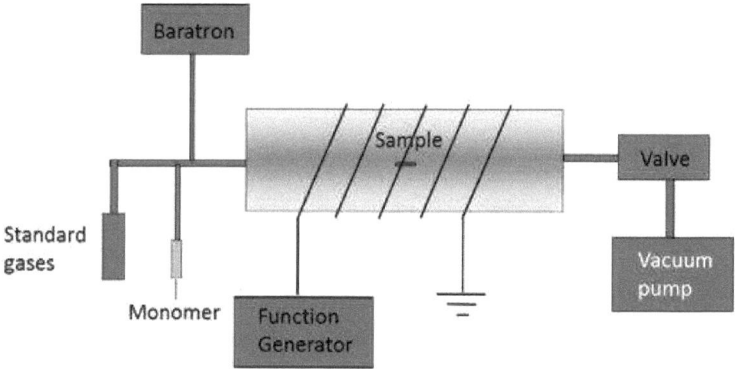

Figure 2-3. Scheme of the plasma chamber used throughout this work.

Prior to plasma polymerization reactors were cleaned for 30 min with plasma containing a mixture of argon and oxygen (flows were 3 sccm and 10 sccm respectively) at 150 W.

Table 2-1 provides details on the plasma depositions. In cases where the parameters varied they will be discussed in the according chapter in the results section of this work.

Table 2-1. Monomer flow rates, input powers, and deposition times.

monomer	flow rate / g·min^{-1}	power / W	time / min
HMDSO	0.037 ± 0.001	30	1/60
AA	0.120 ± 0.003	n.a.	n.a.
Zn(acac)$_2$	0.038 ± 0.004	30	n.a.

Hexamethyldisiloxane (HMDSO) was purchased from Acros (Fisher Scientific, Nidderau, Germany) with 99.99% purity. It was used to increase adhesion of subsequent coatings for all silicon and titanium based substrates via Si-O-Si or Si-O-

Ti bonds.[99] Allylamine was also obtained from Acros with 98% purity, kept in a refrigerator at 4 °C and was brought to room temperature prior to usage. $Zn(acac)_2$ was purchased from Sigma-Aldrich (Steinheim, Germany). All precursors were used without any further purification. Chemical structures are shown in Figure 2-4.

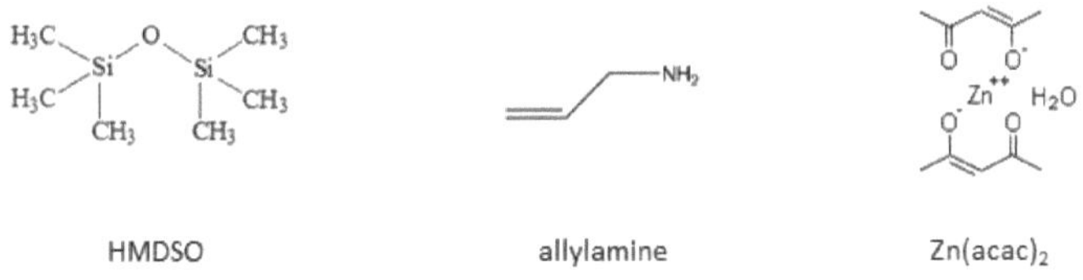

Figure 2-4. Chemical structures of the monomers used throughout this work.

Both HMDSO and allylamine are liquids with a vapor pressure high enough to ensure evaporation by applying vacuum. $Zn(acac)_2$ is a solid with a melting point of 135 °C. It is not going into the vapor phase even at 10^{-6} bar without heating. Hence, a silicon oil bath was used to heat up the flask containing the monomer to 125 °C. Nevertheless, the vapor would resublimate at the inner wall of the glass reactor before reaching the substrate. Therefore, all parts of the glass reactor from the monomer containing flask up to the area where the samples are placed were heated to 125 °C with the help of a heating jacket. This jacket contained thermally isolated heating filaments controlled by a homebuilt temperature controller connected to a temperature sensor in the heating jacket.

2.2 Physical Vapor Deposition

In order to deposit metallic thin films of chromium, silver, or gold onto surfaces physical vapor deposition (PVD) was used. PVD is a procedure in which a small amount of metal or metal alloy is heated in a vacuum chamber to obtain a metal vapor; hence, the method used here is also called evaporative deposition.[123] An Edwards FL 400 (Crawley, England) PVD sctup was used where the heating element consists of a small resistive pan bearing the metal. Upon applying current the receptacle in which the metal is placed heats up. The pressure of the vacuum chamber lies at 10^{-6} bar which together with the heat creates the metal vapor. Deposition rates were kept at 0.6 nm/min. All substrates were cleaned with ethanol prior to PVD. The adhesion of gold films on glass substrates was increased with a 2 nm chromium layer between glass and gold. In order to enhance subsequent polymer adhesion to gold films allyl mercaptane was used in a 50 mM ethanolic solution to create a self-assembled monolayer overnight. This organic monolayer contains double bonds which form reactive species during plasma treatment.[88] It was always applied on gold surfaces before further plasma depositions were performed and its structure is depicted in Figure 2-5. Allyl mercaptane was purchased from Lancaster (Ward Hill, USA).

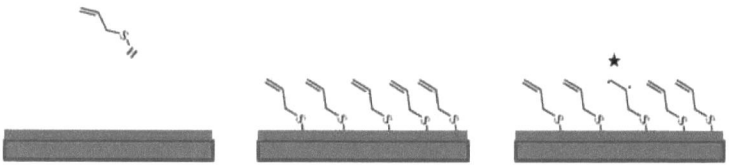

Figure 2-5. Scheme of the formation of a self-assembled monolayer from allyl mercaptane and subsequent activation by plasma (asterix).

2.3 Step Profiling

Step profiling is a method for the determination of film thickness. Polished silicon wafers which were previously cleaned with ethanol were used as a substrate for the deposition of plasma polymer films. After deposition the film was scratched with a steel needle. Since the silicon wafer is harder than steel the depth of the scratch is equivalent to the film thickness. Upon scanning over the scratch with a fine stylus a profile is obtained. Since the stylus is scanning the profile and its position is actuated by a piezo controller the depth of the scratch and therefore the film thickness can be determined. Figure 2-6 shows the basic principle of how a step profiler works.

The equipment used was a KLA Tencor P-16+ (Milpitas, USA) with a scan rate of 50 μm/s at 100 Hz, a needle weight of 1 mg and a tip diameter of about 0.1 μm. Measurements were conducted at least 3-fold for each point of interest.

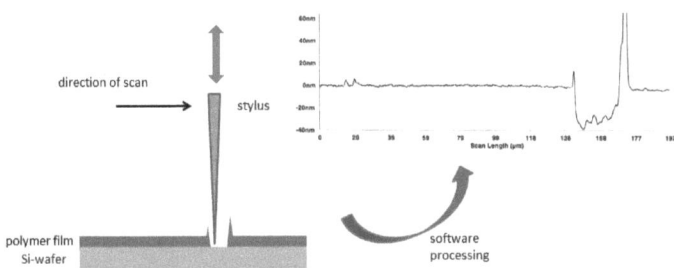

Figure 2-6. Basic mechanism of step profiling. The stylus can move freely in the z-direction and its position is converted into depth information.

2.4 Contact Angle Goniometry

Biological assessment of surfaces and the estimation of their performance can be judged by conducting water contact angle measurements. Mammalian cells prefer a certain range of wettability and 40 to 70° is an appropriate range for cellular adhesion and migration.[124]

Water contact angle goniometry is a method that determines the wettability of a surface by measuring the angle between water and a surface. Usually the measurements are performed in air. This can be done with a sessile drop of water by placing the droplet onto the surface and imaging it from the side. The software provided with the setup determines the angle of the droplet at the point of contact with the surface. If this droplet is not moved or its volume not changed one refers to this procedure as static water contact angle measurement.[125] The angle can change upon movement of this droplet. Advancing contact angles are larger due to retention of the liquid front caused by inhomogeneity of the surface. At the same time receding angles are smaller for the same reason.[126,127] Figure 2-7 shows how the water contact angle is defined in a static measurement.

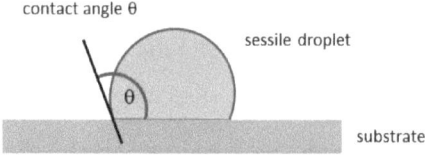

Figure 2-7. Scheme of a static sessile drop water contact angle measurement.

Static water contact angles were determined for the films deposited on various substrates which were cleaned with ethanol prior to surface modification. Static water contact angles are characteristic for the surface modifications and did not depend on the substrate used. A 3 µl droplet of ultrapure water was placed on the sample and the syringe removed to avoid contact with the droplet. The latter is necessary since the droplet and with it the contact angle would otherwise be deformed. The contact angle was determined with a Krüss DSA 10-MK2 (Hamburg, Germany). Measurements were conducted at least 3-fold for each point of interest. The ultrapure water was generated by a water purification system Millipore Milli-Q+ 185 (Molsheim, France) and had a resistivity of 18.2 MΩcm.

2.5 Infrared Spectroscopy

In order to gain information about the chemical structure of the plasma deposits infrared spectroscopy can be utilized. Molecules can absorb electromagnetic radiation and can undergo vibrational changes. These vibrational energies usually lie in the range of infrared light and are specific for chemical bonds and functional groups.[128] The energy is absorbed and the chemical bond is excited from its vibrational ground state to a higher quantum state. For two chemical bonds associated with a central atom various vibrational modes are possible like symmetric and asymmetric stretching as well as bending vibrations like scissoring and rocking, wagging and twisting. Infrared spectra are usually measured and recorded by observing the absorption of infrared light compared to a reference.

2.5.1 Infrared Reflection Absorption Spectroscopy

One way of receiving structural information of thin films in the nanometer range is the utilization of infrared reflection absorption spectroscopy (IRRAS). A reflecting gold film (80 nm) is coated with the film and illuminated with infrared light with an incident angle of about 5°. The infrared light passes the film and is reflected by the gold layer after which it passes the film again. The reflected light is recorded with a detector which is cooled with liquid nitrogen to increase the signal-to-noise ratio (SNR). As a reference an uncoated gold coated sample was used. The technique is very sensitive and the surfaces need to be free from contaminants. This is ensured by cleaning the samples with ethanol prior to deposition of polymer films. Figure 2-8 shows the incident and detected infrared beam during interaction with a polymer film on gold.

Figure 2-8. Schematic representation of the mode of function in IRRAS.

Microscopy glass slides were coated with 80 nm of gold using PVD as described in section 2.2. Glass slides were purchased from Menzel (Braunschweig, Germany). The setup used was a Thermo Scientific Nicolet Magna IR 850 (Nidderau, Germany). The detector was cooled with liquid nitrogen 15 min prior to and during measurements.

2.5.2 Attenuated Total Reflection Infrared Spectroscopy

The aforementioned method relies on the reflection of an underlying gold layer. Attenuated total reflection infrared spectroscopy (ATR) is a method that is not depending on such a reflecting surface. A crystal with high refractive index, on which the sample is placed on, is illuminated with infrared light. The infrared light is reflected from the inner surface of the crystal as it provides an internal reflection element. Upon reflection the electric field of the infrared light creates an evanescent wave on the outer surface of the crystal. This evanescent wave interacts with the sample in close proximity. To ensure close contact between sample and crystal (to increase signal-to-noise ratio) light pressure is applied on the specimen. After internal reflection the beam is guided to a detector which is cooled with liquid nitrogen (again to increase signal-to-noise ratio).

A Thermo Scientific Nicolet FT-IR 730 (Nidderau, Germany) was used here. The detector was cooled with liquid nitrogen 15 min prior to and during measurements.

2.6 Immobilization of Biomolecules

Wet chemical treatment was carried out to increase the cell-adhesive properties of ppAA. A straight forward approach for amino group coupling of biomolecules is the use of a short linker bearing an epoxy group at either end.[115] The epoxide ring structure lowers its energy upon opening with the help of a nucleophile. In this case the lone pair of nitrogen can provide sufficient nucleophilicity to do so. Diethylene glycol diglycidyl ether (DEGDGE) was used as a linker molecule and was dissolved in 50 mM $NaHCO_3$ at a concentration of 5% v/v. The samples were fully immersed in 1 ml of the solution and incubated at room temperature for 30 min without stirring.[2]

Figure 2-9 shows that the linker molecule reacts with only one end, although in principle loop structures may easily be formed on the surface. However, the high concentration of the linker kinetically favors the reaction with only one end. Subsequently the samples were rinsed with 50 mM $NaHCO_3$ at least three times and were then placed in a solution of the same buffer containing 0.5 mg/ml RGD (linear or cyclic form) or fibronectin. Reaction was carried out for 30 min at room

---

[2] No stirring was used since kinetic measurements were performed as described later. This procedure did not allow for stirring such that this step was not conducted in all experiments to ensure comparability.

temperature. After coupling the samples were rinsed with the buffer three times and subsequently washed with ethanol.

The linear and cyclic form of RGD was purchased from Anaspec (Fremont, USA) whereas fibronectin, $NaHCO_3$, and DEGDGE were purchased from Sigma (Deisenhofen, Germany).

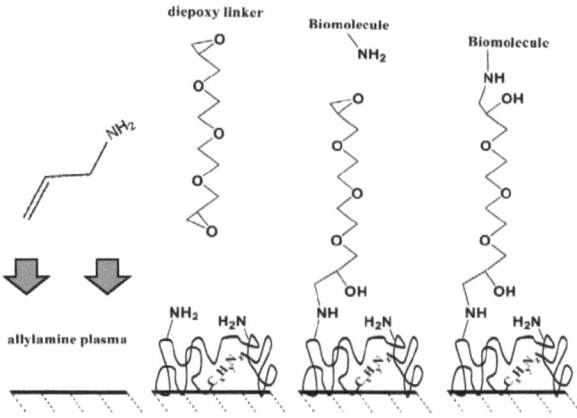

Figure 2-9. Scheme of the deposition and biofunctionalization of ppAA.

2.7 Surface Plasmon Resonance

Surface plasmon resonance (SPR) is the physical process of electron oscillations at the interface of a conductive material, e.g. gold and silver, and a dielectric, e.g. polymer films. These oscillations can be induced with electromagnetic radiation such as visible light. Upon a certain incident angle the electric field of the light, or rather the z-component, induces an oscillation which is attenuated and localized; hence it is called evanescent field, or evanescent wave.[129,130] In a thin metal film the plasmon interacts with a dielectric material opposite of the incident beam such as a polymer film. The incident angle of the beam at which a plasmon occurs strongly depends on the thickness and refractive index of the polymer film. Figure 2-10 shows the working principle during an SPR measurement.

A detector monitors the light reflected from the gold surface. During the plasmon resonance the intensity of reflected light decreases to zero because all the energy of the light is consumed to create the plasmon field. Due to the fact that only the z-component can contribute to plasmon oscillations[3] x- and y-components would make it more difficult to observe the resonance phenomenon. Therefore, the laser light is send through polarizers that will only allow z-polarized light to pass through.

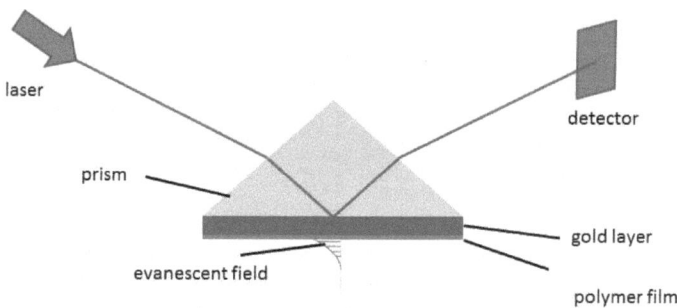

Figure 2-10. Basic scheme of the working principle of SPR measurements.

The properties (thickness and refractive index) of the deposited layer on the gold are influenced by attachment on and detachment from the surface. As a result, the method allows for the observation of adsorption processes on the deposited layer. Figure 2-11 shows this for two SPR measurements. The red line was recorded prior and the black line after an adsorption process.

After determining the angle at which the plasmon occurs, this signal can be followed via software that controls a goniometer. This goniometer ensures that the incident angle of the light is changed according to the shift of the resonance angle. This way, a time evolved process that causes the resonance angle to change can be observed.

---

[3] x- or y-components may very well cause electron oscillations but these never screen outside of the gold layer and therefore are not influenced by any adsorbed layer in contact with the gold. Exactly these adsorbed layers (e.g. polymer films, proteins) are supposed to be investigated by this technique.

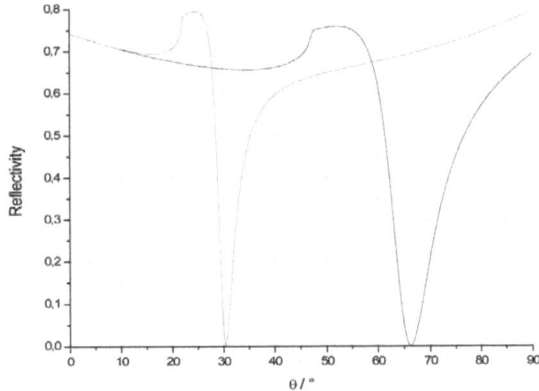

Figure 2-11. Example of two plasmon resonances recorded. The signal on the right (black line) shows a higher resonance angle resulting from a change in thickness and refractive index of an adsorbed layer.

For kinetic SPR measurements Hellma LaSFN9 glass (Müllheim, Germany) previously cleaned with ethanol was coated with 47 nm of thermally evaporated gold and was subsequently immersed in a 50 mM ethanolic solution of allyl mercaptane overnight. A HeNe laser with 633 nm was used to excite the plasmon. Flow cells made from PTFE were attached on the coated side of the sample. This offers the ability to do wet chemistry while the measurement is conducted. Solutions and rinsing buffers were carefully injected with a syringe. The wet chemical procedure was performed as described in section 2.6.

2.8 Inductively Coupled Plasma Optical Emission Spectroscopy

Optical emission spectroscopy was utilized to determine the metal concentrations released from the metal containing multilayer system. By transferring energy to a solution containing the material of interest, the elements are excited and can emit light of a distinct wavelength. Energy transfer, absorption, and emission processes can be described by a so-called Jablonski diagram.

In this case the excitation energy is provided by hot (up to 10,000 K) argon plasma ignited by inductive coupling to a high frequency generator. This configuration and process is called inductively coupled plasma optical emission spectroscopy (ICP-OES).[131,132] The solution is led into the plasma as an aerosol created by a nebulizer and all liquid components vaporize immediately upon the high temperatures in the plasma. The residual substances and particles melt and vaporize as well and undergo excitation processes with respect to their electron structure and configuration. The emission lines can be spectrally analyzed by a monochromator and a detector. The resulting signal is compared with standard solutions of known concentrations. This way, concentrations can be measured down to the ng/ml range. A basic scheme of the working principle is shown in Figure 2-12.

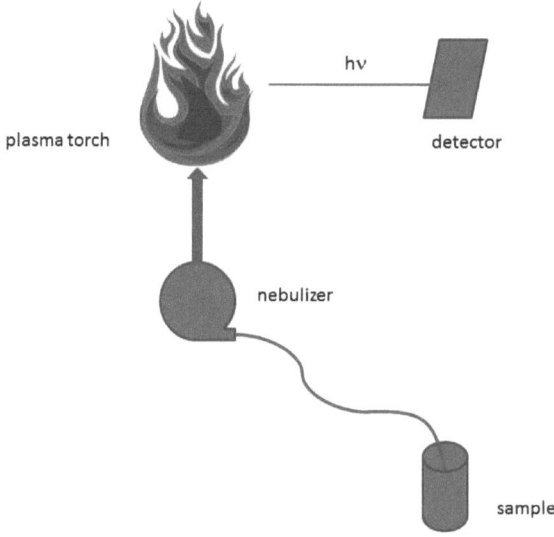

Figure 2-12. Basic principle of the light detection from a plasma torch fed with an aerosol of a liquid sample. The light emitted can be analyzed in a detector.

Round glass cover slips (thickness 1.5 mm, borosilicate, VWR, Darmstadt, Germany) with a diameter of 13 mm were used as substrates. They were coated and subsequently immersed in 1 ml of ultrapure water. The supernatant was collected at a certain point in time and the sample was reincubated with 1 ml of fresh ultrapure water for an additional point in time. Each sample was then diluted 4-fold with ultrapure water to obtain sufficient volume for the analysis. A

Horiba Jobin Yvon Activa M (Unterhaching, Germany) was run at an input power of 1250 W and an argon flow of 12 l/min.

2.9 Scanning Electron Microscopy

Scanning electron microscopy (SEM) is able to resolve micro- and nano-structures by scanning a sample with a focused electron beam. A high resolution can be reached compared to standard optical microscopy due to the short wavelength of the electrons used. According to Abbé's law the maximum resolution is given by half of the wavelength. The electrons are emitted from a heated electron gun and are accelerated by an electric field up to several keV. By using apertures and magnetic lenses the beam can be focused onto desired areas of the sample. By absorption and scattering of the primary electron beam further particles are formed such as secondary electrons and X-rays. The secondary electrons can be captured in a detector and by scanning over the surface with the primary electron beam each point of the sample observed contributes to an image. By using low voltage (a few keV) charging and the creation of artifacts can be avoided.[133] Secondary electrons have only small energies well below 50 eV and only leave the sample if they were created close to the surface of the sample. Therefore, SEM images represent the outer structure and topography of the sample.

For cross section views polished Si-wafer were used as a substrate for the deposition of the films. These wafers were then cracked by placing a thin wire underneath them and applying pressure to both ends of the substrate. Thus, two pieces were obtained with a clean cross section at the breaking edge. A Carl Zeiss Gemini 1530 (Oberkochen, Germany) was used to image the samples normal to the breaking edge at vacuum conditions of 10^{-9} bar.

2.10 Energy-Dispersive X-Ray Spectroscopy

As already mentioned in the previous section, interaction of energetic electrons with matter creates secondary electrons as well as X-rays. Whereas the secondary electrons originate from the surface of the sample, the X-rays come from deeper layers of the sample. The energy they bear is strongly depending on the elemental composition of the analyzed specimen. The technique is commonly known as energy-dispersive X-ray spectroscopy (EDX). The principle behind elemental detection is the Auger effect. It states that upon impact of electrons with sufficient energy, electrons from inner shells of an element can be released. The lack of an electron is compensated by an electron from the next higher energy level. The surplus energy that is set free upon the transition is released in the form of X-rays. By detecting the energy one can determine the element in which the X-ray was created. Since the initial primary electron beam is localized a lateral resolution in the nanometer range is obtained. Therefore, scanning over the same breaking edge as described in the previous section gives information about the elemental composition of the multilayers as they are designed in this work. Due to the similarity in the primary source of energy (primary electron beam) SEM and EDX are often combined in one instrument. Sample preparation and equipment was the same as described in section 2.9.

2.11 X-Ray Photoelectron Spectroscopy

Another method to investigate chemical composition of surfaces is X-ray photoelectron spectroscopy (XPS). Upon exposure to X-rays from a magnesium or aluminum source the elements in the surface release electrons. Part of the energy of the X-ray E_x is used to overcome binding energy of the electron E_b. The kinetic energy of this electron E_k is detected by an analyzer. Since the energy of the X-ray is known the binding energy can be determined:

$$E_b = E_x - (E_k + \varphi) \qquad 1.3$$

Because the detector is an energy analyzer that accelerates the electrons the work function of that analyzer needs to be considered to compensate for that acceleration when measuring E_k.[134,135] The determination of the binding energy is the basis of the identification of elements in the surface. The number of electrons counted is proportional to the quantity of the respective element.[136] Figure 2-13 shows the basic principle of XPS electron transition upon irradiation with X-rays.

The sampling depth is about 10 nm due to the relatively low mean free path of the electron. This makes XPS a very surface sensitive technique.[137] Nevertheless, it makes it also very demanding since impurities and contaminations are easily detected by this technique. To prevent premature interaction of X-rays with any elements the whole procedure is carried out at 10^{-9} bar. Due to the loss of electrons from the surface the sample may charge. This can either be prevented by using electrically conducting or semi-conducting materials, covering the samples with a contacting metal grid, or by applying low energy electrons on the surface with an electron flood gun.[138,139]

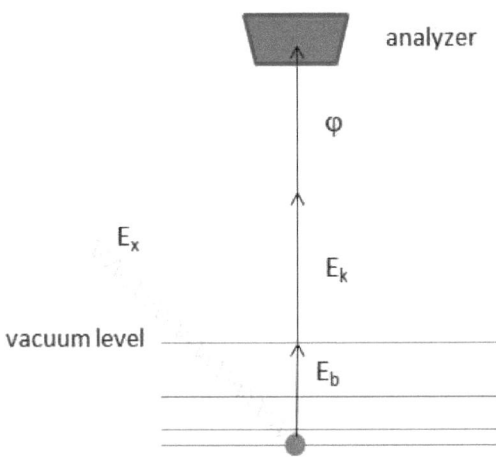

Figure 2-13. Principle of XPS with electron transition. In the bound state the electron absorbs energy and bears a specific kinetic energy as a free electron.

In order to estimate the amount of primary amines in plasma polymerized allylamine (ppAA) a selective wet chemical derivatization procedure with 4-(trifluoromethyl) benzaldehyde[140,141] was conducted as depicted in Figure 2-14. Since XPS is very sensitive to fluorine and no fluorine is contained in the samples otherwise this procedure can help to estimate the amount of primary amines on the surface. Samples were placed in a closed glass container on 2 mm glass beads. A droplet of 4-(trifluoromethyl) benzaldehyde was then placed on the bottom of the container and the samples were exposed to the vapor at 45 °C for 3 h. By reaction with the derivatization agent a single primary amine bears three fluorine atoms. Quantification of fluorine by XPS therefore reveals the amount of primary amines.

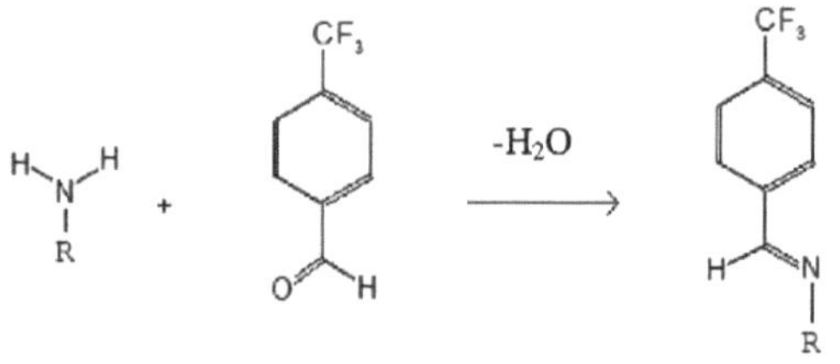

Figure 2-14. Derivatization reaction of primary amines from a ppAA plasma deposit (R) with 4-(trifluoromethyl) benzaldehyde.

The setup used was a VG EscaLab 3 MKII and experiments were conducted by the École Polytechnique Montréal. The setup is equipped with a MgKα source with 1253.6 eV X-rays. The screened surface was about 2 x 3 mm. The step size for the energy scan was 1 eV. The Shirley algorithm was used for background correction of the element signals. Charging effect corrections were made with the C1s peak being set to 285.0 eV. Polished Si-wafers were used as substrates and ensured minimal charging of the surfaces.

2.12 Overview over the Surface Analysis Techniques

Table 2-2 gives an overview over the surface analysis tools used throughout the work and includes lateral resolution as well as analysis depth of the respective technique.

Table 2-2. Overview over the surface analysis techniques used in this work.

name	information obtained	lateral resolution	analysis depth
Step Profiling	film thickness	µm	whole film
Contact Angle Goniometry	wettability	mm	interface
Infrared Reflection Absorption Spectroscopy	chemical composition	mm	whole film
Attenuated Total Reflection-FTIR	chemical composition	mm	whole film + substrate
Surface Plasmon Resonance	optical thickness (adsorption)	mm	nm - µm
Scanning Electron Microscopy	topography	nm	interface
Energy-Dispersive X-Ray Spectroscopy	elemental composition	nm - µm	µm
X-Ray Photoelectron Spectroscopy	elemental composition	µm	~ 10 nm

2.13 Bacterial Assays

Since it is the aim of this work to create antimicrobial surfaces that minimize bacterial attachment, colonization, and biofilm formation on surfaces it is a crucial step to expose the coatings to bacterial suspensions. The design of the test method depends on the mode of action of the antimicrobial surface. Whereas contact killing surfaces bearing quaternary ammonium compounds or antimicrobial peptides do not act on bacteria in the surrounding medium, the surfaces studied in this work actively release antimicrobial agents and affect bacteria in the solution. The latter case is easier to observe since merely the solution in contact with the surface needs to be analyzed and standard methods have been developed for such cases. *E. coli K12* (Top10, Invitrogen, Darmstadt, Germany) as well as two clinical strains *S. aureus MSSA 476* (cultured by Dr. Jodi Lindsay Geraldine Mulley, Department of Biology & Biochemistry, University of Bath, Bath, BA2 7AY, United Kingdom, St Georges University, London, UK) and *Pseudomonas aeruginosa PAO1* (same source) were utilized. Due to safety issues all experiments with *S. aureus* or *P. aeruginosa* were conducted at the Pathology Department of the University of Cologne and the Hygiene Department of the University Medical School in Mainz.

The simplest method is to measure the optical density of the supernatant since bacteria are sufficiently large to scatter light. The more bacteria are present in a solution the more light is scattered and the more optically dense this solution will be. By measuring the light going through such a solution and comparing it to standards with known number of bacteria one can quantify the bacteria in the sample. Usually light with $\lambda = 600$ nm is used since most of the culture media do not absorb in that region; hence the measured quantity is called OD_{600}. However, no distinction between living and dead bacteria can be made but by comparing the result to a control sample with no antimicrobial activity the results may offer a first overview over the feasibility of the surface modifications. The photometer used for the measurements was a Thermo Fischer Scientific Genesys 10 UV (Braunschweig, Germany). For such experiments 1 ml of a bacterial suspension with 10^6 colony forming units (CFU) in 1/10 Luria Bertani (LB) medium (Carl Roth, Karlsruhe, Germany) were incubated in polystyrene petri dishes (Nalge Nunc International, Rochester, USA) with a diameter of 3.5 cm overnight at 37 °C and 95% relative humidity.

To obtain accurate numbers the supernatant was plated out on agar plates in various dilutions and grown overnight at 37 °C and 95% relative humidity. Since every vital bacterium will form a new colony, the number of CFU in the original supernatant can be determined by counting colonies on the agar plate and considering the dilution factor. Experiments involving *E. coli* were performed by Sabine Pütz (Max Planck Institute for Polymer Research, Mainz, Germany). Experiments with silicone urethral catheters and *S. aureus* were performed by Peter Cierniak (Pathology Department, University of Cologne, Germany).

For assessing the longevity of antimicrobial properties the method was changed in such way that no dilution series of the supernatant was made and no counting of CFU was conducted. Plating merely served as a way to judge how long the bactericidal effect lasted. PTFE (round discs of 13 mm diameter, Cadillac Plastic, Swindon, England) was chosen since it is clinically relevant and a material that is most challenging with regard to wet chemical procedures.[142] After cleaning with ethanol and deposition of the films described in the previous chapter 10 µl of a suspension of 10^6 CFU/ml in 1/10 LB medium were incubated on the samples at 37 °C and 95% relative humidity overnight. The suspension was then transferred to an agar plate and grown over night to see if bacteria survived. The original sample was again incubated with a fresh bacterial suspension and this process was repeated until bacterial growth was not further inhibited. The time up to the point of bacterial growth was then considered as the longevity of antimicrobial properties of a given surface. These experiments were conducted by Nina Dohm (University Medical School, Mainz, Germany).

Another method that allows for observation of bacteria on the surface and assessing their viability is a so-called live/dead staining. The kit used was Invitrogen Live/Dead BacLight (Darmstadt, Germany) which contains two nucleic stains, a green fluorescent dye (Syto 9) and a red fluorescing dye (propidium iodide) These stains differ in their ability to penetrate the membrane of vital bacterial cells. Only Syto 9 (not charged) can penetrate the membrane of both live and dead bacteria whereas propidium iodide (positively charged) will only penetrate that of dead bacteria. Therefore, dead bacteria fluoresce in red and vital bacteria fluoresce in green. Samples were inoculated with 10^6 CFU/ml of bacteria in 1/10 LB medium and incubated at 37 °C and 95% relative humidity overnight. For one sample 3 µl of each BacLight component were mixed with 1 ml of 0.85% NaCl in sterile water (autoclaved). After removing the supernatant from the samples and gentle washing with 1 ml of PBS this solution was incubated with the samples for 15 min. By observation with a microscope (Carl Zeiss Axio Observer, Oberkochen, Germany) with the use of appropriate filters (485 and 530 nm) it is possible to image both dyes simultaneously. This is done by overlaying the images with the software provided with the microscope. Fabrics made from woven (Sefar, Heiden, Switzerland) and non-woven (Freudenberg, Weinheim, Germany) polyethylene terephthalate (PET) was used for the Live/Dead staining assays.

Since plasma itself is sterilizing due to highly reactive species like radicals, ions, free electrons and UV light[143] no further sterilization of the samples was conducted with the exception of wet chemically treated samples. The latter as well as all the equipment used for handling the samples (tweezers, glass vials, et cetera) were washed with 70% ethanol.

2.14 Cell Microscopy

Since fast wound healing helps to prevent infections and implant failure it is important to assess the cell behavior on the surface coatings. Furthermore, antimicrobial properties may also result in cytotoxicity and it is the aim here to combine both bactericidal as well as cell-adhesive properties. For this purpose, cell morphology, number, and surface coverage of cells were investigated. *Fibroblasts NIH 3T3* (DSMZ, Braunschweig, Germany) were grown overnight on the various films deposited on glass cover slips. These glass samples were 13 mm in diameter and were placed in 24-well plates. The number of cells seeded was 10,000 per sample in 1 ml of Dulbecco's Modified Eagle Medium (DMEM) (Invitrogen, Darmstadt, Germany) and growth conditions were kept at 37 °C and 95% relative humidity. After incubation for 24 h the samples were observed through a microscope. The microscope used was a phase contrast Olympus IX70 (Hamburg, Germany).

Human umbilical vein endothelial cells (HUVECs) purchased from Pan Biotech (Aidenbach, Germany) were also used since they are clinically relevant especially for the assessment of coatings for vascular grafts. The cells were cultivated in Endopan3 medium (Pan Biotech) at 37 °C and 95% relative humidity. The samples, made of PTFE or titanium, were seeded with 10,000 cells per sample in a 24-well plate. Cell attachment was monitored after 24 h and 3 d. For the visualization of the attached cells the samples were rinsed with PBS and subsequently incubated in FCS-free DMEM containing 1 mg/ml calcein-AM. Calcein-AM is the acetoxymethylester of calcein. It prohibits the calcium chelating effect of calcein and thus enables diffusion through the cell membrane. Within a vital cell the ester is cleaved by esterases and calcein is formed. Upon chelating calcium it can fluoresce at 515 nm (excited at 495 nm). After 30 min of incubation at 37 °C samples were washed with medium. Micrographs were obtained using standard FITC filters. At least three different, representative spots from each sample were photographed. The surface coverage of attached cells was quantified using ImageJ software (NIH, Bethesda, USA) and expressed as percentage of total observed area. For this purpose, three pictures of each of the tested surfaces were converted into an RGB stacked image (red, green, and blue channels). Using the green channel the threshold for particle recognition was adjusted manually and the picture converted into a binary image (black and white). The software could then determine the area covered by cells as well as their number. Sterilization of the samples was performed as described in the previous section.

Throughout this work all substrates were thoroughly cleaned several times with HPLC grade ethanol in an ultrasound sonicator for 15 min. The only exceptions were silicon wafers and glass cover slips since they are too thin and brittle to endure sonication. Nevertheless, they were cleaned by washing with ethanol before conducting any further treatment.

3 Cell-Adhesive Coatings with ppAA

The following chapter will address immobilization of fibronectin on ppAA[113] and provides the basis of the work in chapter 5. The aim was to create coatings that enhance cell attachment on materials for biomedical applications. It was shown previously that plasma polymerized allylamine (ppAA) exhibits cell-adhesive properties because of the amount of amino functional groups covering the resulting film.[144-146]

It was furthermore shown that ppAA can be used to immobilize biomolecules by physisorption[108,147,148] as well as chemisorption[97,149]. The fundamental principle of utilizing biomolecules to prompt a certain behavior of mammalian cells is applied here by grafting fibronectin onto ppAA and subsequently observing the cell adhesion and morphology of HUVECs on modified PTFE samples. A basic scheme of the utilized layer system is shown in Figure 3-1.

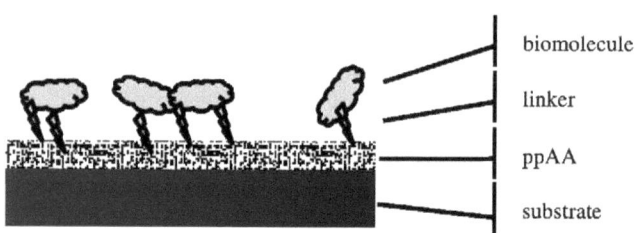

Figure 3-1. Scheme of the layer system discussed in chapter 3.

3.1 Analysis of Cell-Adhesive Coatings

Many traditional polymers are susceptible to plasma and will easily bond to a plasma polymer film[4] thus ensuring stable coatings.[145,150] Nevertheless, upon immersion in aqueous solutions plasma polymer films can delaminate from some

---

[4] Often an argon and/or oxygen plasma is used to create reactive species on the substrate to further enhance stability, especially in cases were a 'soft' (low input power) plasma deposition is chosen for the sake of retaining monomer structure and functionality.

substrates, e.g. silicon based substrates like Si-wafers or glass as well as metallic substrates like gold, silver or titanium. Delamination can be overcome by using plasma polymerized hexamethyldisiloxane (ppHMDSO) as an adhesion layer between the substrate and ppAA.[151,152] This works well for substrates from titanium and silicon based materials. The ppHMDSO deposit can attach to the surface with Ti-O-Si or Si-O-Si bonds while providing organic moieties at which a subsequent plasma deposit can attach to.[99]

3.1.1 Film Adhesion Improvement with ppHMDSO

In order to obtain thin ppHMDSO films small pieces (few millimeters in dimension) of polished Si-wafers were evenly distributed throughout the reactor chamber in an area of 16 by 6 cm. Figure 3-2 shows the film thickness according to the position in the reactor for a 5 s deposition of HMDSO (0.100 mbar) at 40 W CW and an oxygen carrier gas flow of 10 sccm.

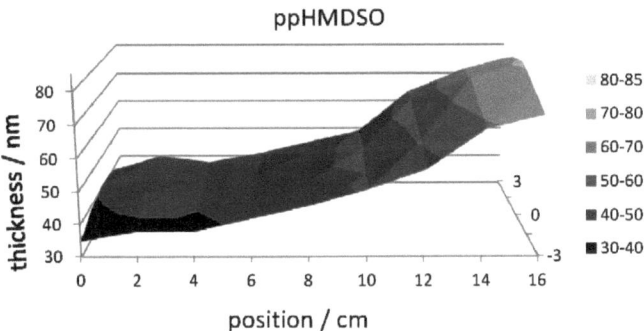

Figure 3-2. Film thickness distribution throughout the reactor after 5 s of HMDSO deposition. The precursor inlet is situated on the left whereas the vacuum pump is placed on the right.

Film thickness increased from the precursor inlet towards the pump (distal end). It is not yet understood why this is the case. However, it is also the direction of precursor flow and the reactor chamber forms a bottleneck towards the distal end lowering the diameter of the chamber from 10 cm down to 3 cm. This may well lead to an accumulation of monomers therefore causing a higher partial pressure of the monomer in that region.

Nevertheless, the variation in film thickness did not influence the stability of the subsequent plasma polymer layers. In all following experiments with silicone based substrates a short deposition of 2 s was used to create ppHMDSO films of 10 to 20 nm in thickness. These could then further be activated by a short oxygen plasma treatment with low input power (30 W). As has been shown previously this does not alter the general organic nature of the ppHMDSO but increases film stability, adhesion, and reactivity.[151-153]

Figure 3-3 shows IRRAS spectra of ppHMDSO thin films on gold before (green) and after (red) exposure to oxygen plasma for two seconds at 30 W CW at a flow of 10 sccm.

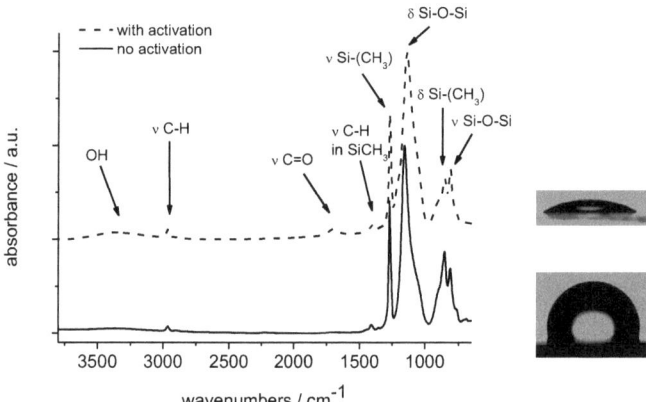

Figure 3-3. IRRAS spectra of ppHMDSO with (red) and without (green) subsequent plasma activation with oxygen. Water contact angles are shown by the side view images of a water droplet on the surfaces (also refer to Table 3-2 in section 3.1.2).

Both spectra demonstrates the retention of the organic character of the thin films by showing specific signals (see Figure 3-3 for details and assignments) which were in accordance to what has been reported before.[151] This was also true for the wettability. While ppHMDSO had a water contact angle of about 100°, the contact angle of ppHMDSO treated with 'soft' (30 W) oxygen plasma decreased to 35° (see Table 3-2 in section 3.1.3 for details). It has been shown before that hydrophilic ppHMDSO itself leads to adhesion of proteins[154] and the increase of adhesion and proliferation of fibroblasts on ppHMDSO coatings with and without oxygen activation was studied.[155] Here, ppHMDSO merely serves as an adhesion promoting layer between substrates and additional plasma polymer layers.

3.1.2 Analysis of ppAA Films

In this work ppHMDSO merely serves as an adhesion mediator between silicon and titanium based substrates. The subsequent deposition of ppAA builds the basis of the cell-adhesion experiments.

In order to investigate deposition rates small Si-wafer pieces (few millimeters in dimension) were placed in the reactor and were subsequently coated with ppAA at 0.150 mbar for 1 min and 40 W CW. As shown in Figure 3-4 deposition rates were higher towards the bottleneck at the distal end of the precursor inlet, which is in agreement to what has been observed for ppHMDSO.

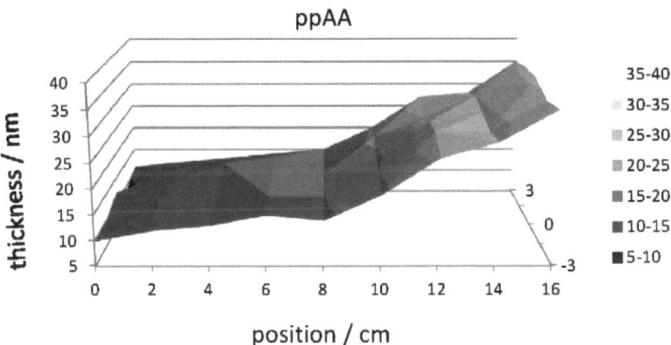

Figure 3-4. Film thickness distribution throughout the reactor after 1 min of allylamine deposition. The monomer is located on the left whereas the pump is placed on the right.

Additionally, for IR analysis three distinct positions inside the reactor were chosen; close to the precursor inlet, in the middle, and close to the pump (see Figure 3-5). Also, three different input powers were used in order to vary the composition of ppAA. Table 3-1 furthermore gives detailed assignment on the various signals in accordance with the literature.[97,144,156-158]

All spectra in Figure 3-5 were normalized to the signal at 1650 cm^{-1} and were background corrected manually to ensure comparability between them. The signals at 1450-1370 cm^{-1} (seen as two signals at the right shoulder of the main peak at 1650 cm^{-1}) as well as the signals at 3000-2800 cm^{-1} serve as a reference for the loss of functionality of the plasma polymer film because they originate from aliphatic structures of carbon and hydrogen. Relative changes in intensity between hydrocarbon signals and the signal at 1650 cm^{-1} (mainly attributed to the primary amine) are therefore discussed.

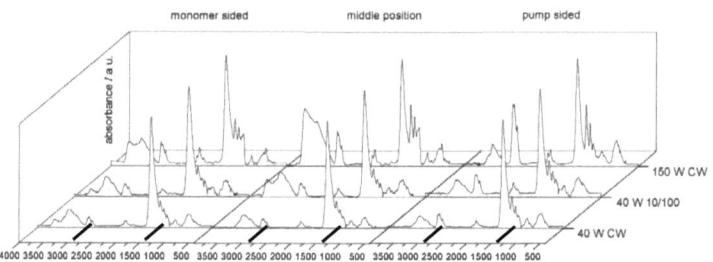

Figure 3-5. IRRAS spectra of three different positions in the reactor and three different input powers (y-axis). Spectra were normalized to the strongest signal at 1650 cm-1. The signals marked (black lines) are representative for aliphatic hydrocarbon structures and are discussed in the context of loss of amino functionality. All spectra were background corrected manually to ensure comparability.

Table 3-1. IR signals of ppAA

wavenumber / cm^{-1}	chemical structure
3340	=N-H stretch, N-H, and O-H stretch
3280	NH_2 stretch
3190	R_3N-H stretch
3000-2800	R_3C-H stretch
2870	=C-H stretch
2240	C≡N and C≡C stretch
2180	N=C=N stretch
1650	NH_2 bend, , C=N stretch
1450	CH_2 in plane
1370	CH_2 and CH_3 deformation
850	R_3N-H bend

The 150 W high input power films exhibit especially strong signals attributed to hydrocarbon structures and are therefore unsuitable for the following experiments which rely on retention of amino functionality.

The films deposited in pulsed mode at 40 W 10/100 show lower signals for aliphatic structures but such low input power films are prone to loss of film structures upon immersion in liquids or autoclaving.[159,160] The amount of crosslinking within these plasma polymer films is usually too low. They are therefore unsuitable for further experiments.

The best compromise between stability and retention of the monomer structure seems to be reached at 40 W CW. All aliphatic hydrocarbon structures were relatively low compared to films deposited at high input powers. Delamination or loss of film structure was not observed except for what is already known in literature.[159] With regard to the position in the reactor there seem to be minor differences. At the distal end (closer to the vacuum pump) higher signals for aliphatic structures were observed. However, these differences were relatively small and may also originate from baseline corrections[5]. A broad homogeneity throughout the reactor is favorable because many samples can thus be coated simultaneously.

Due to the fact that plasma deposition is mostly substrate independent[118,161] it is possible to deposit different plasma polymer films on top of each other without compromising integrity of the sub-layers of plasma polymer films. Therefore, deposition of ppAA on top of ppHMDSO is easily achieved and can be observed via infrared techniques as well as simply measuring water contact angles (see Table 3-2 in section 3.1.3 for details). Both results are presented in Figure 3-6 where infrared spectra of the ppAA, ppHMDSO, and the bilayer as well as contact angles are shown. Corresponding signals are highlighted by colored arrows revealing the presence of both films. Assignments were given in Figure 3-3 and Table 3-1. It is worth noting that although film thickness is in the same order of magnitude (10 nm for ppHMDSO and 20 nm for ppAA) the signal for ppAA is relatively weak compared to ppHMDSO.

Not all substrates need ppHMDSO as an adhesion enhancing interstitial layer. Classical polymeric substrates like PTFE can be plasma modified[162] and ppAA was directly deposited onto these substrates after 10 min pretreatment with oxygen plasma.

---

[5] Baseline correction was merely conducted to make the spectra comparable. Throughout this work baseline correction was avoided wherever possible.

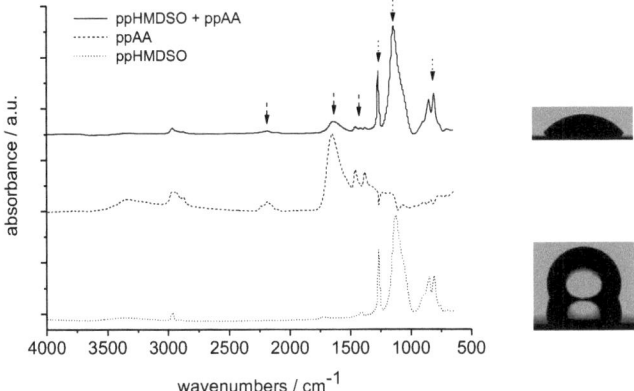

Figure 3-6. Normalized IRRAS spectra of ppAA on ppHMDSO. Colored arrows mark signals of the respective films (details are shown in Figure 3-3 and Table 3-1). The according water contact angles are shown next to the spectra (also refer to Table 3-2 in section 3.1.3).

Figure 3-7 shows a spectrum of ppAA on top of PTFE as well as spectra of the substrate and ppAA only.[6] The two main signals at 1200 and 1140 cm^{-1} originated from stretching vibrations of fluorocarbons and they are characteristic for PTFE. The signals assigned to ppAA were present on coated PTFE samples. Also, static water contact angles change drastically upon deposition of ppAA on PTFE. The latter exhibits contact angles of 110° whereas plasma polymerization of allylamine lowered the water contact angle to 35° (see Table 3-2 in section 3.1.3 for details).

---

[6] The ppAA spectrum was obtained with IRRAS. The PTFE and the PTFE + ppAA spectra were obtained with ATR. Peak areas or height are therefore not directly comparable. Nevertheless, peak positions are.

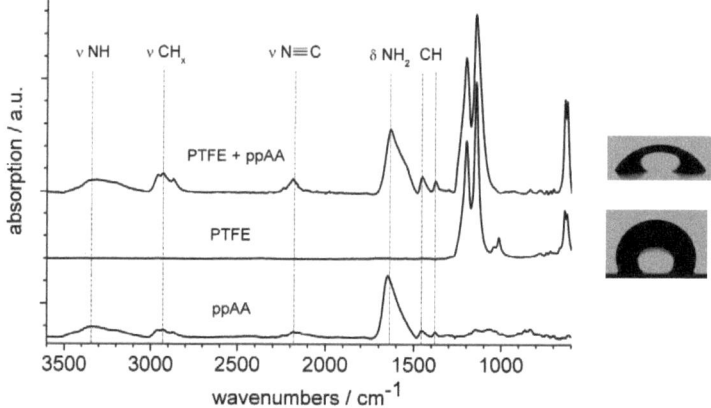

Figure 3-7. ATR-FTIR spectra of ppAA on PTFE. Also shown are spectra of PTFE and ppAA only. Distinct signals of ppAA are highlighted (see Table 3-1 for details). The according water contact angles are shown next to the spectra (also refer to Table 3-2 in this section).

Besides infrared absorption spectroscopy techniques and water contact angle measurements, X-ray photoelectron spectroscopy (XPS) was utilized to give further information on the composition of ppAA and the retention of amino functional groups. Elemental scan data will be given later on in section 6.1 but it is worth noting here that the elemental composition of ppAA was close to that of the monomer. Almost 20% of the films composition was made of nitrogen compared to 25% in the precursor. Utilizing 4-(trifluoromethyl) benzaldehyde which can be used for selective derivatization of primary amines[140,141] has shown a high retention of monomer structure. It can be concluded that 28.7% ± 0.3% of the total nitrogen concentration originated from primary amines and about 10% of the carbon atoms bore primary amine moieties.

3.1.3 Biofunctionalization of ppAA

The purpose of ppAA deposits was to create a cell-friendly environment and to produce cell-adhesive surface modifications. Therefore, fibronectin and the adhesion sequence RGD in its linear and cyclic form were chosen to be immobilized on ppAA coated substrates and devices.

A binding chemistry that has previously shown to be applicable for traditional polymers was utilized for the grafting of fibronectin to ppAA (see section 2.6).[115] Diethylene glycol diglycidyl ether (DEGDGE) is a short polyethylene glycol like linker molecule with two terminal epoxide rings and is used to covalently bind to amino functional groups. It can react with either end and in principle may form loop structures on the ppAA surface. Since loop structures do not contribute to further binding to amino functional groups of biomolecules they need to be avoided using a large excess of this diepoxy linker. This way the single sided reaction is kinetically favored. DEGDGE is readily soluble in aqueous solutions such as 50 mM $NaHCO_3$. Here, a concentration of 5% v/v in a 1 ml per sample was utilized. The alkaline solution with a pH of 9 enhances the nucleophilic attack on a carbon of the epoxide ring, most likely on the sterically less demanding terminal carbon. The electron lone pair of either amino functional group can therefore readily react with the ring structure leaving pendant epoxide moieties which either hydrolyze or react with other amino functional groups. The latter will be favored if the subsequent addition of the biomolecule solution is started shortly after binding the linker.

Here, the reaction time of the linker was 30 min in order to minimize hydrolysation. Figure 3-8 shows a kinetic SPR measurement. Figure 3-8 a) represents the procedure used for all experiments involving this grafting chemistry. After adding DEGDGE in buffer the signal increased strongly (from 60.8 to 63.2°) which is mainly due to the optical properties (high refractive index) of the solution and does not indicate any binding event yet. During washing with buffer after 30 min the signal decreased below the starting point to 60.6° (overall decrease at this point 0.2°). Plasma polymer films including ppAA are known to release material upon immersion in aqueous solution[141,159] but various other effects take place simultaneously; plasma polymer films swell in solution therefore changing their refractive indices as well as their thickness but as already mentioned they also release material. All of these processes affect SPR signals and are hard to distinguish. Nevertheless, a plateau is reached after a few minutes and upon incubation with fibronectin solution the signal increased again to 61.5° indicating an adsorption process. Up to this point, physi- and chemisorption cannot be distinguished though most of the adsorbed mass remained bound after rinsing with buffer.

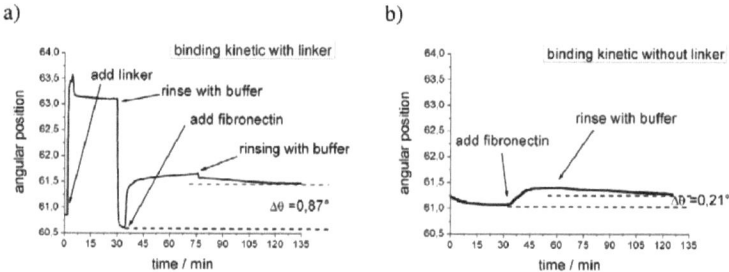

Figure 3-8. Kinetic SPR measurements of fibronectin adsorption to ppAA a) with and b) without DEGDGE. The angular position (corresponding to adsorption or desorption processes) is plotted against the time.

In contrast to the results presented in Figure 3-8 a) the spectrum looks different when no linker is used. This is shown in Figure 3-8 b) and the result implies that the process relied only on physisorption. During incubation with buffer for 30 min[7] the signal decreased by 0.2° (swelling of the film and loss of plasma polymer material as discussed above) and stabilized. After incubation with fibronectin solution the signal increased but not to the extent seen in the case of utilization of the linker. After rinsing with buffer, the signal remained well below that of the linker modified samples. The overall increase in angular position was only 0.21° (no linker used) compared to 0.87° (linker used). The results indicate that the utilization of the linker molecule caused chemisorption compared to physisorption that was always present in either case.

It is likely that due to its size of 520 kDa[163] fibronectin binds to many sites on the surface and reacts with many linker molecules. In principle this may cause the

---

[7] This step serves as a substitute for incubation with the linker solution and makes the results comparable.

protein to lose its functionality due to the loss of the natural tertiary structure. Nevertheless, it has been shown before that fibronectin retains its biofunction even after denaturation.[111] This is because of the relatively small binding domain consisting mainly of the RGD sequence which the protein can display even upon denaturation.[164] In principle a quantification of such biomolecules is possible but involves techniques such as the Western Blot[165] which were not conducted during this work.

Figure 3-9 shows IRRAS spectra of ppAA and the subsequent layers after wet chemical treatment. For comparison, spectra of fibronectin as well as of the DEGDGE linker were added. There are various signals found in the fingerprint region after incubating ppAA with DEGDGE and fibronectin. Among them the anti-symmetric and symmetric epoxide ring deformation at 908 cm^{-1} and 842 cm^{-1} respectively.[166] The most prominent signal is the C-O-C stretching vibration from ethylene glycol structures at about 1150 cm^{-1} which were not present in unmodified ppAA.[167]

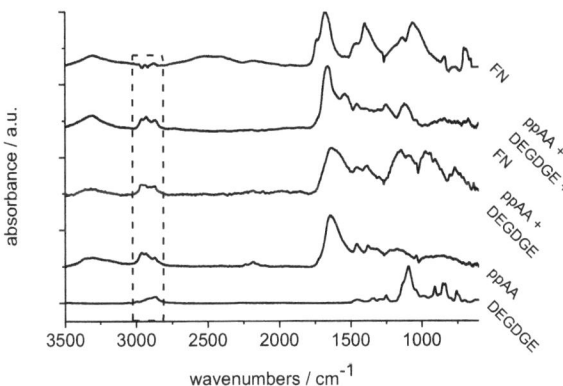

Figure 3-9. IRRAS spectra of the wet chemically modified plasma deposit. The black box marks the detailed section shown in Figure 3-10.

A clearer picture can be derived from the hydrocarbon signals between 3100-2800 cm^{-1} which can be seen in greater detail in Figure 3-10. Spectra of

DEGDGE and ppAA are shown in the lower part whereas the wet chemically modified films are shown in the upper portion of the figure. At 3060 cm^{-1} the anti-symmetric stretching of CH$_2$ in the epoxide ring structure can be seen.[166] It is not a prominent feature but was present in case of DEGDGE and ppAA + DEGDGE but did not show after binding fibronectin. This suggests that either fibronectin saturated all the linker molecules and/or the linker hydrolyzed. That the former is more likely was shown by the kinetic SPR measurements (see above). There are further signals at 3000 cm^{-1} and 2860 cm^{-1} which are attributed to hydrocarbon vibrations that were found in both spectra. Nevertheless, they were enhanced after wet chemical treatment which is pointing to formation of new bonds after ring opening of the epoxide structure.

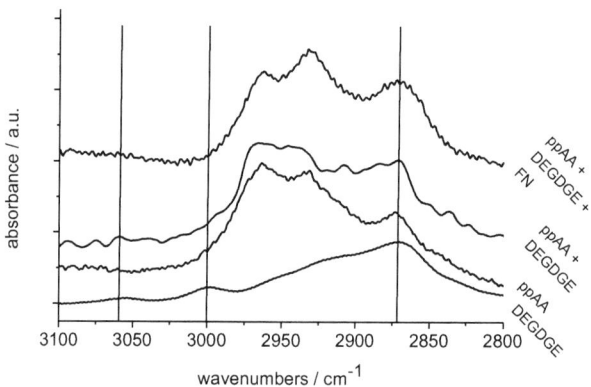

Figure 3-10. Detailed section from Figure 3-9. The distinct signals are highlighted with black lines.

Table 3-2 shows static water contact angles of various surfaces at the different stages of preparation. It is worth noting that wettability of PTFE did not change after 10 min of oxygen plasma treatment at 100 W CW. Gold and silicon based substrates like glass and Si-wafers exhibited typically low contact angles and the software of the measurement system failed to determine reliable numbers (they are therefore merely denoted <10° in Table 3-2). All plasma deposits developed a specific static water contact angle independent of the substrate.[8] Water contact angles after all wet chemical modifications were in a comparable range of 30 to 35° which makes distinction between them impossible by this method only.

Table 3-2. Water contact angles of various surfaces used in the context of cell adhesion on ppAA.

surface	water contact angle / °
PTFE	110 ± 2
PTFE with activation	108 ± 4
glass	< 10
gold	< 10
silicon wafer	< 10
ppHMDSO	98 ± 6
ppHMDSO with activation	35 ± 5
ppAA	52 ± 4
DEGDGE on ppAA	32 ± 4
fibronectin	33 ± 4
L-RGD	33 ± 4
C-RGD	33 ± 4

3.1.4 Cell Adhesion on Biofunctionalized ppAA Films

For cell experiments HUVECs were chosen because they provide a clinically relevant biological system, especially for applications involving vascular grafts made from PTFE. It is desired to have fast attachment and growth of endothelial cells on such devices. Endothelialization of PTFE would prevent platelet adhesion and activation of the coagulation cascade. This coagulation together with foreign body

---

[8] Note that this may very well not be true for dynamic water contact angles, which were not measured here.

reactions are the main cause for thrombosis and implant failure of vascular grafts.[168-170]

Cell adhesion was determined with microscopy after seeding HUVECs on treated and untreated PTFE discs. Figure 3-11 shows micrographs of calcein stained HUVECs after 24 h and 3 d of incubation. It is evident that PTFE did not allow cells to adhere to the surface due to its hydrophobic and chemically inert nature. This is implied by the spherical morphology of the cells minimizing cell-surface interaction. Also, oxygen plasma treated PTFE did not improve cell adhesion. This was not expected but became obvious after measuring static water contact angles (see Table 3-2 above) which proved that these samples remained hydrophobic. Nevertheless, all subsequent samples were cleaned by oxygen plasma treatment prior to ppAA deposition.

Deposition of ppAA on PTFE allowed HUVECs to adhere to the surface which is seen by the enhanced cell-surface interaction. The cells changed their morphology and adopted a well-spread shape. There was a difference in adhesion between the 24 h and 3 d results revealing a better attachment after the longer period. This is important to notice as it shows that ppAA remained stable on the chemically inert PTFE material and that ppAA furthermore remained active as an adhesion promoting surface modification after several days. After wet chemical treatment with DEGDGE cell adhesion was reduced which is due to hydrolyzation of the linker taking the form of PEG-chains[9] on the surface which are known to have anti-fouling properties.[171] It can be expected that the same chemistry may be exploited for anti-fouling strategies in other applications by using PEG-chains with higher molecular weight exhibiting even less cell adhesion.

---

[9] In this case PEG (polyethylene glycol) is misleading and due to the short length of the linker it should not be called PEG but merely shows a similar structure.

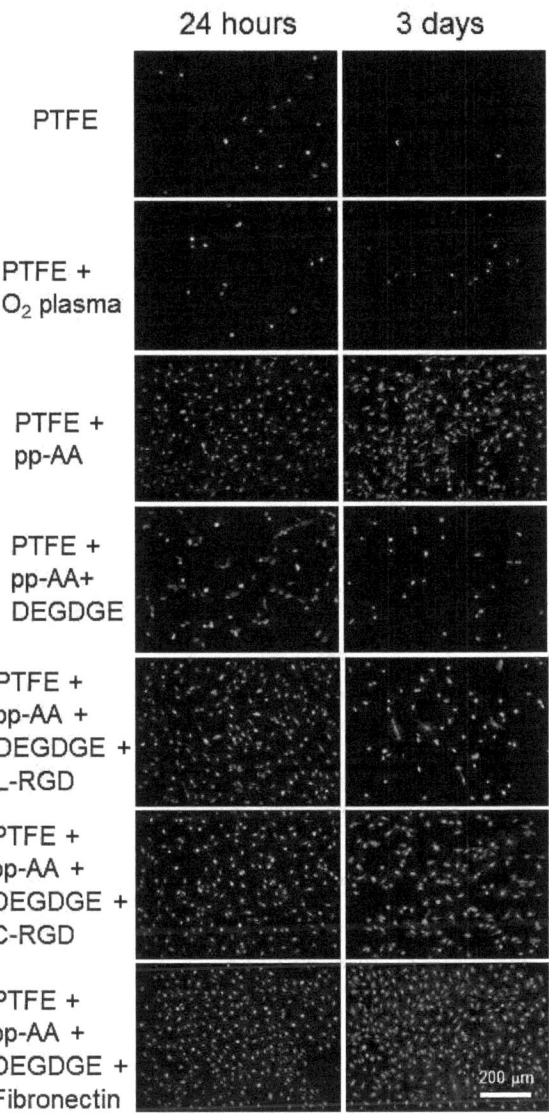

Figure 3-11. Calcein stained HUVECs after 24 h and 3 d on PTFE samples with various surface modifications based on ppAA deposits.

RGD modified surfaces, whether using the linear or the cyclic form, promoted cell adhesion.[164,172,173] Initial attachment seemed to be fast and there was no difference between cell morphology after 24 h and 3 d for both the linear or the cyclic peptide. Also, cell adhesion did not exceed that of ppAA after 3 d.

Fibronectin immobilized on the surface strongly enhanced cell adhesion. Cell numbers and morphology implied good attachment and after 24 h the sample already showed an almost confluent coverage.

Further data can be gained, by converting the images into binary pictures (black and white) and evaluating the number and surface coverage of the cells. This was done with an automated algorithm provided with the software ImageJ. Figure 3-12 shows the cell numbers observed on the specified samples and the results correspond to what has already been discussed. Coating PTFE with ppAA as well as with the biomolecules increased the number of cells compared to the control whereas grafting DEGDGE showed reduced cell numbers.

It is worth noting that the amount of cells did not change between 24 h and 3 d except for both forms of RGD where cell numbers in fact decreased. The reason for this is unknown at this point. A possible explanation may include a fast boost of cell attachment and growth in the initial phase followed by an inhibitory effect in proliferation. Nevertheless, this can only be proven with appropriate methods including proliferation tests (e.g. Alamar Blue assay) which were not conducted here. Again, fibronectin showed a strong effect on the cell behavior and strongly increased cell growth compared to all other surface modifications or the control.

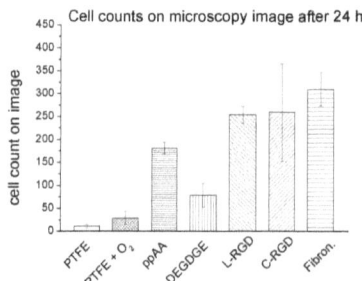 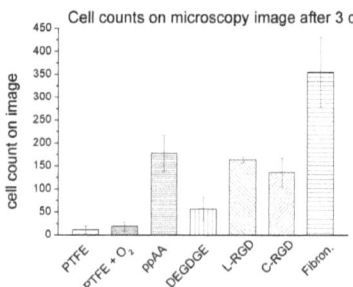

Figure 3-12. Cell counts evaluated from pictures shown in Figure 3-11.

The coverage of the surface by cells can also be quantified using the ImageJ software. Figure 3-13 shows the results which are in good agreement to the microscopic pictures. Compared to linear RGD the cell coverage on the cyclic form was slightly increased after 3 d. In the literature it is also claimed that the cyclic form of RGD offers better cell adhesion compared to the linear form.[174] This is thought to originate from the natural arrangement of RGD within the tertiary structure of fibronectin which forms a loop domain.[107,175] Nevertheless, the error is relatively large here and as shown above cell numbers and morphology did not differ between the two RGD structures. It is worth mentioning that the coverage is almost confluent in the case of fibronectin coated samples whereas it is almost zero for the PTFE control. Immobilization of fibronectin therefore offers the most promising cell adhesion strategy.

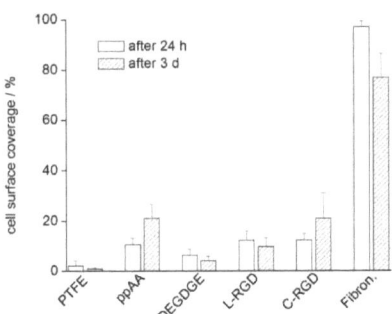

Figure 3-13. Cell coverage derived from pictures shown in Figure 3-11.

4 Antimicrobial Coatings with ppZn(acac)$_2$

The aim of this chapter is to discuss antimicrobial coatings from plasma polymerization of metal-organic precursors among which ppZn(acac)$_2$ seems favorable due to its high vapor pressure compared to other metal-organic compounds.

4.1 Analysis of ppZn(acac)$_2$ Films

Analysis of the zinc containing plasma polymers included investigation of the distribution of film thickness in the reactor as well as on confined samples like petri dishes. Also static water contact angles were observed and the zinc content in the films was determined. Further structural analysis and investigation of chemical composition is given in the following chapters.

4.1.1 Plasma Deposition of ppZn(acac)$_2$

The zinc-organic monomer zinc acetylacetonate (Zn(acac)$_2$) was heated to 125 °C with a silicon oil bath in order to create sufficient amounts of vapor needed for a glow discharge. In addition to heating the monomer containing flask, the parts of the reactor that fed the vapor into the chamber were heated as well. Heating filaments regulated by a temperature controller heated the reactor head to 125 °C from the outside to minimize resublimation at the glass wall. Nevertheless, glass is not a good thermal conductor and temperature might differ between outside, where the temperature sensor for the feedback loop is positioned, and the inside, where resublimation can occur. Figure 4-1 shows the temperature setpoint compared to the actual temperature at the inner surface of the glass wall. The temperature slightly remained below the setpoint but was considered sufficiently high to prevent resublimation. The straight line represents the bisector between x- and y-axis at which perfect heat convection would exist.

Air and vacuum have a low thermal conductivity and therefore the temperature at the sample position was also measured. There was a large difference between the setpoint and actual temperature, e.g. at 125 °C being set in the controller the temperature at the sample position merely reached about 90 °C. This is acceptable since the monomer is supposed to polymerize and to deposit onto the substrate at this position. High temperatures over a prolonged period of time could reduce deposition rates or might even decompose the precursor.

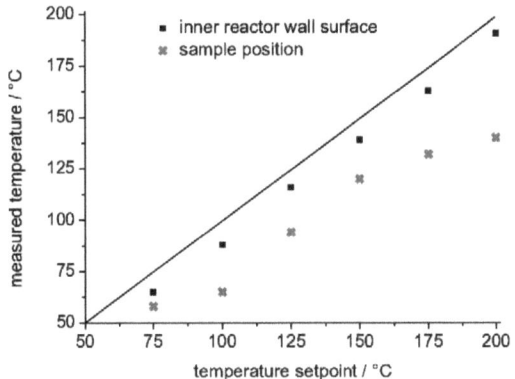

Figure 4-1. Temperatures in the reactor (at the inner reactor wall surface and at the position where the samples were placed) versus the setpoint of the temperature controller. The straight line represents the angle bisector between x- and y-axis (a theoretical perfect heat convection).

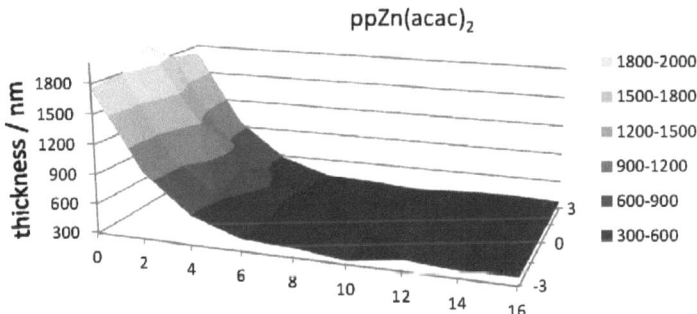

Figure 4-2. Film thickness distribution throughout the reactor after 30 min of ppZn(acac)2 deposition.

The film thickness distribution inside the reactor is depicted in Figure 4-2. In contrast to the results shown in chapter 3.1 where HMDSO and allylamine gave thicker films towards the distal end of the reactor close to the cold trap, the behavior of $Zn(acac)_2$ is the reverse. Due to the lower vapor pressure compared to that of HMDSO or allylamine, $Zn(acac)_2$ can resublimate inside the reactor. Although the heating system is preventing this, it does not extend over the entire reactor but ends right at the point of highest deposition rates. Beyond that position the lack of external heat will cause the partial pressure of the precursor to decrease drastically leading to slower film growth. It is worth mentioning that the difference in film thickness can reach 500% for $ppZn(acac)_2$ within a few centimeters. Therefore, throughout this work sample position for following experiments was restricted to the position of highest deposition rate.

Limited access to confined spaces of samples is a known issue of plasma polymerization.[176] Therefore it is interesting to investigate how deposition rates alter for a confined sample since antimicrobial efficacy was tested in 3.5 cm diameter polystyrene petri dishes with a rim height of 1 cm. Figure 4-3 shows the film thickness distribution on a radius of such a petri dish. Along the first 2 mm from the rim, film thickness was lower compared to more accessible areas in the center of the petri dish. However, the thickness was considered high enough not to influence any experimental results. Also, reduced film thickness was limited to the first 2 mm from the rim only. It therefore comprised only a small portion of the petri dish.

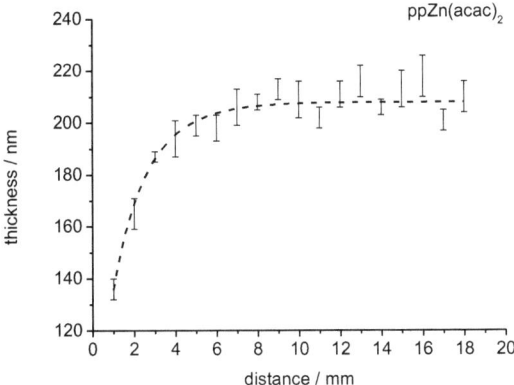

Figure 4-3. Film thickness of ppZn(acac)2 within a petri dish of 3.5 mm diameter going from the rim to the center.

Table 4-1 gives a short summary of static water contact angles relevant for this chapter. It is worth noting that there is a clear distinction between hydrophobic polymeric materials like polystyrene or polyethylene terephthalate (PET) and ppZn(acac)$_2$. The latter is more hydrophilic and readily absorbs water during the measurement so that a starting contact angle can only be approximated. The fact that this plasma polymer tends to change its static water contact angle within seconds reveals its ability to absorb water without completely dissolving. The antimicrobial properties depend on the dissolution and diffusion of zinc from those water absorbing deposits.[95]

Table 4-1. Water contact angles of various surfaces used in the context of antimicrobial coatings with ppZn(acac)2.

surface	water contact angle / °
polystyrene	88 ± 2
woven PET	78 ± 3
ppZn(acac)$_2$	~ 60

4.1.2 Zinc Content in ppZn(acac)$_2$ Films

Plasma polymerization merely affects the surface without altering the bulk properties of the substrate. But it is possible to influence the structure of the resulting plasma polymer film by further exposure to a glow discharge.[118] This is due to ion bombardment and results in sputtering, crosslinking, and chemical restructuring. Hence, it is worth to investigate the zinc content of ppZn(acac)$_2$ in dependence of the polymerization duration or film thickness. This is possible by dissolving the zinc embedded in the deposit with concentrated nitric acid and quantifying the zinc content with ICP-OES. Table 4-2 shows the zinc content per volume element of the films for varying ppZn(acac)$_2$ film thickness. About 275 mg/cm^3 of zinc were found independent of the film thickness. It is therefore evenly distributed in the plasma deposit and the total amount of zinc is easily controlled by controlling the film thickness.

Table 4-2. Zinc content per volume element of ppZn(acac)2 films.

thickness ppZn(acac)$_2$ / nm	zinc content / mg/cm^3
100	282 ± 19
250	270 ± 11
500	265 ± 16

4.2 Antimicrobial Efficacy of ppZn(acac)$_2$ Films

Antimicrobial properties of the presented surfaces rely on zinc originating from the coating and going into solution. The focus of this section is on the release of zinc as well as the potential application of ppZn(acac)$_2$ for medical devices.

4.2.1 Optical Density of Bacterial Suspensions

Antimicrobial properties were investigated by incubating 4 ml of a bacterial suspension with 10^6 CFU/ml of either *S. aureus* or *P. aeruginosa* in 1/10 LB medium overnight on coated and uncoated polystyrene petri dishes. After collecting the supernatant a dilution series was made in order to measure OD_{600} for quick estimation of the applicability of the approach.

Figure 4-4 shows the results of this experiment. Growth of *S. aureus* was inhibited by ppZn(acac)$_2$ whereas that of *P. aeruginosa* was not and even exceeded the bacterial growth on the control. OD_{600} measurements can only give general information about bacterial behavior but it clearly reveals that *P. aeruginosa* was not stressed by ppZn(acac)$_2$. Other approaches must be found to overcome this limitation. Resistance to oxidative stress caused by zinc[177] has been observed before for gram-negative bacteria like *P. aeruginosa*[178] whereas gram-positive bacteria seem to be more susceptible. This is most likely due to their lack of an outer membrane. A solution to overcome the limitations of zinc based antimicrobial approaches will be presented in chapter 6.

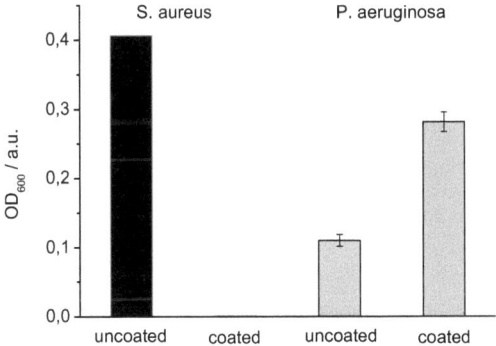

Figure 4-4. Optical density of bacterial suspensions of S. aureus and P. aeruginosa after incubation in uncoated and coated polystyrene petri dishes overnight.

4.2.2 Coating Woven Fabrics with ppZn(acac)$_2$

Besides coating petri dishes, woven fabrics made from PET fibers (Sefar, Switzerland) which are used as filter material for medical applications were treated with ppZn(acac)$_2$. Figure 4-5 shows photographs of the uncoated and coated material. A change in color is visible and does not seem to be completely homogenous. The reason for this was the varying deposition rate inside the reactor as shown in Figure 4-2 in section 4.1.1. Also, the fabric was attached in upright position in the reactor to achieve coating on both sides simultaneously. To position the fabric in such fashion it was attached to the reactor with an adhesive tape. These spots were therefore left uncoated and were not used for any experiments. Appropriate samples were cut out with scissors (cleaned with ethanol prior to usage). A more homogeneous coating would be reached by laying the sample flat on a sample dish. However, this procedure is only suitable for one-sided treatment and would require a second treatment on the other side. Since smaller pieces were cut out from the sample the inhomogeneous coverage was accepted.

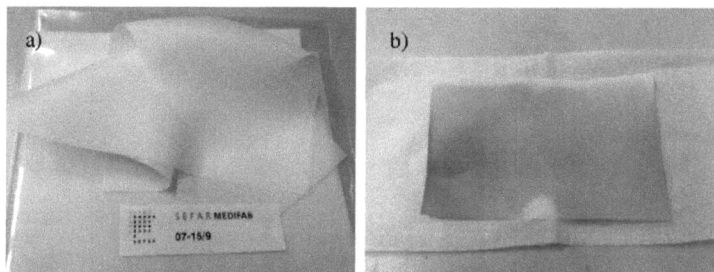

Figure 4-5. Photographic images of a) uncoated and b) ppZn(acac)2 coated woven PET from Sefar.

Energy-dispersive X-ray spectroscopy (EDX) spectra were taken from both samples (Figure 4-6) revealing that the control only contained carbon and oxygen (next to hydrogen which is not detectable by EDX). The coated material additionally contained zinc as expected. A quantitative analysis was not conducted due to relatively low signal to noise ratio (SNR).

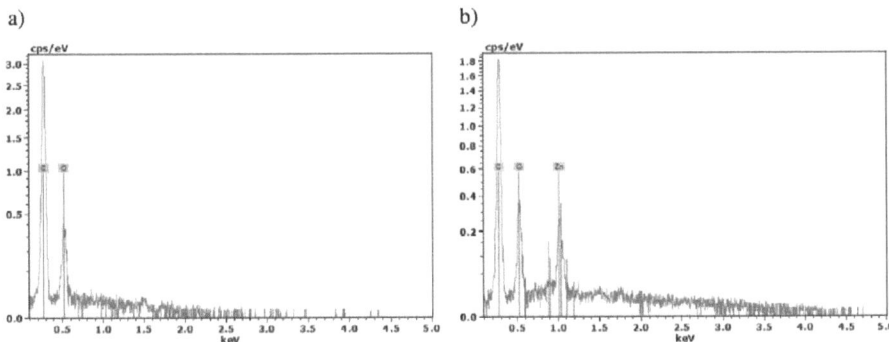

Figure 4-6. EDX spectra of the woven PET samples shown in Figure 4-5.

Both the untreated and treated PET samples (4 cm^2) were incubated overnight with 4 ml of a suspension containing 10^6 CFU/ml of *S. aureus* in 1/10 LB medium. Subsequently, the samples were stained with Baclight Live/Dead (Invitrogen, Germany) which stains vital bacteria in green and dead bacteria in red. These dyes always give a background staining. Only stained spots or particles should therefore be considered since they represent stained bacteria. Figure 4-7 shows micrographs of the samples after performing the assay revealing that all bacteria are vital on untreated PET whereas a coating of 500 nm of ppZn(acac)$_2$ kills most of the bacteria. Cytocompatibility and cytotoxicity is not yet discussed here and is postponed to sections 5.3.3 and 5.3.4.

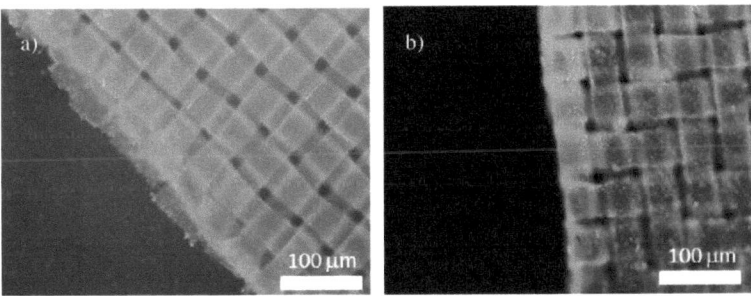

Figure 4-7. Live/Dead staining of S. aureus on a) uncoated and b) ppZn(acac)2 coated woven PET shown in Figure 4-5.

5   Bilayered Coatings with ppAA and ppZn(acac)$_2$

The aim of the work described in this chapter was to investigate the amount of zinc released from the surface of ppZn(acac)$_2$ and bilayers of ppZn(acac)$_2$ and ppAA. Furthermore, a method of controlling the release is presented. Because of the importance for medical applications the behavior of bacteria and mammalian cells was observed. A basic scheme of the layer system discussed in this chapter can be seen in Figure 5-1 and is thought to have potential for medical application and industrial exploitation.[179]

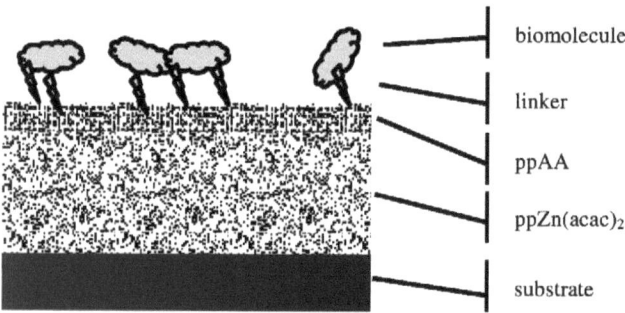

Figure 5-1. Scheme of the layer system discussed in chapter 5.

5.1   Plasma Deposition of ppAA on ppZn(acac)$_2$

In order to control the release of zinc from the ppZn(acac)$_2$ films an additional top-layer, consisting of another plasma polymer film, was deposited and serves as a diffusion barrier for zinc. This barrier needs to be stable, hydrophilic, and swellable to enable diffusion of aqueous solutions. For this purpose ppAA was chosen which furthermore provides the basis for additional wet chemical modifications. The deposition and film properties of ppAA have already been discussed in section 3.1.2 and the focus here lies on the combination of ppAA and ppZn(acac)$_2$ plasma polymer deposits.

As already discussed in section 4.1.1 plasma polymerization has limitations in confined areas. This is also true for ppAA and because bacterial testing again included coating polystyrene petri dishes the deposition behavior of ppAA on such a sample was investigated. As shown in Figure 5-2 the first 2 mm from the rim of the petri dish deposition rates were reduced about 5 to 10 nm. Nevertheless, the majority of the coated area is not influenced and such effects are therefore not taken into account. Errors are relatively large since various batches were considered. Repeatability is reduced due to the short deposition times of allylamine (here 60 s) and the very thin films.

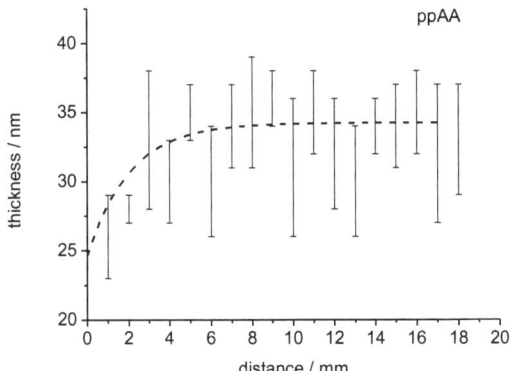

Figure 5-2. Film thickness of ppAA inside a petri dish of 3.5 cm in diameter (from the rim to the center).

Also, the amount of zinc within a volume element of ppZn(acac)$_2$ does not change with the thickness of ppAA on top as can be derived from data shown in Table 5-1.

Table 5-1. Zinc content per volume element of ppZn(acac)2 films modified with top-layers of ppAA in varying thickness

thickness ppAA / nm	zinc content / mg/cm^3
100	270 ± 11
250	261 ± 15
500	275 ± 8

IRRAS spectra were taken from ppZn(acac)$_2$ and ppAA films as well as from bilayers of ppAA in varying thickness on top of a 500 nm ppZn(acac)$_2$ film. The zinc containing layer was relatively thick here in order to obtain a large reservoir for zinc. This is thought to provide increased longevity of antimicrobial properties compared to thinner films. Figure 5-3 shows the spectra of the films. Obtaining signals that reveal both layers simultaneously is challenging due to overlapping absorption peaks. Nevertheless, O-H and N-H bands were detectable at around 3300 cm^{-1} which are attributed to ppAA or water adsorbed on the film. For thicker films of ppAA in the range of 100 nm the signal at 1650 cm^{-1} (see Table 3-1 in section 3.1.2) occured as a shoulder of the ppZn(acac)$_2$ signal at 1595 cm^{-1} which originates from carbonyl in the zinc organic complex.[180] Further signals in ppZn(acac)$_2$ at 1535 cm^{-1} (C=C stretching mode), around 2960 cm^{-1} (aliphatic hydrocarbon modes), and 1390 cm^{-1} (CH$_3$ stretching vibration)[180] can be identified but to a large extent overlap with signals from ppAA.

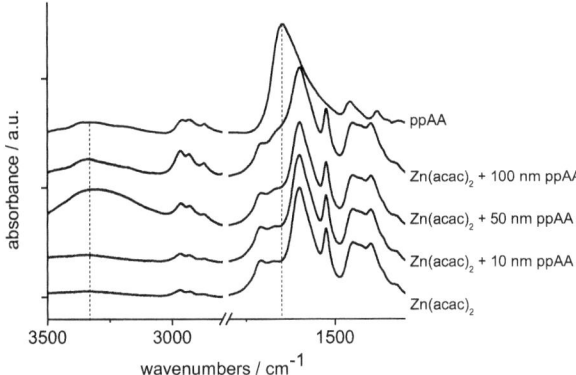

Figure 5-3. IRRAS spectra of ppAA films of varying thickness on ppZn(acac)2. Distinct differences in spectra are highlighted by dashed lines.

Static water contact angles were measured of various surfaces relevant for this chapter and are given in Table 5-2. Each surface modification exhibited a characteristic and substrate independent water contact angle.

Table 5-2. Water contact angles of various surfaces used in the context of bilayered coatings from ppZn(acac)2 and ppAA.

surface	water contact angle / °
titanium	49 ± 2
polystyrene	88 ± 2
glass	< 10
ppZn(acac)$_2$	~ 60
ppAA	52 ± 4
DEGDGE on ppAA	32 ± 4
fibronectin	33 ± 4

5.2 Release from Bilayered Coatings with ppAA on ppZn(acac)$_2$

An additional plasma polymer coating of ppAA could provide a method for controlling the zinc release from the ppZn(acac)$_2$ sub-layer. In order to prove this ICP-OES measurements were performed. The supernatant was collected from samples that were incubated with 4 ml of ultrapure water. Substrates were polystyrene petri dishes that were coated with 500 nm of ppZn(acac)$_2$ and a top-layer of ppAA with varying thickness.

Figure 5-4 shows the change in concentration of zinc in the supernatant over time. Zinc is released from the coating and accumulates in the liquid phase. A film consisting of ppAA on top of the ppZn(acac)$_2$ deposit reduces this release. The extent of this reduction is controlled by the thickness of the top-layer. Each point in time was acquired by collecting 0.5 ml from the supernatant. This amount was considered sufficiently small so that the diffusion of zinc was not disturbed. This seems reasonable considering the small absolute amount of zinc compared to the relatively large reservoir of the supernatant. Nevertheless, later experimental designs were changed to completely collect the supernatant and to reincubate with fresh ultrapure water at each point in time. This is thought to resemble the biological situation where a permanent flow of fluids is present. Instead of an accumulation of zinc in the supernatant one expects a decrease in zinc concentration over time (this will be confirmed in chapter 6.2).

However, Figure 5-4 shows the amount of zinc released from the bilayer over the course of 1 to 120 h after incubation with distilled water. This amount is easily controlled by the thickness of the barrier layer. Without a barrier layer the concentration of zinc increased quickly within the first few hours and continued to rise over the course of 5 d (up to 3 µg/cm^2). By deposition of 14 nm of ppAA as a barrier layer the zinc concentration in the first hours was only half of that compared to the monolayer of ppZn(acac)$_2$. Also the increase of zinc over the following days was much lower. For very thick barrier layers like 140 nm the zinc concentration remained extremely low (below 1 µg/cm^2) over the course of 5 d. Depending on the size of the coated device the amount of zinc released from such layers may not be sufficient for antimicrobial applications. Errors enlarged towards the end of the experiment. Although the petri dishes were stored in a bowl which was lined with wet paper tissue to ensure high humidity unequal evaporation of the solvent is a possible explanation.

In an application one could adjust the thickness of ppAA to meet the requirements considering the size of the device as well as the amount of fluid in the tissue or the flow of bodily fluids around the implant. Both, the thickness of ppZn(acac)$_2$ and the thickness of ppAA can be altered so that the amount of zinc released for a given amount of time and volume matches the medical requirements.

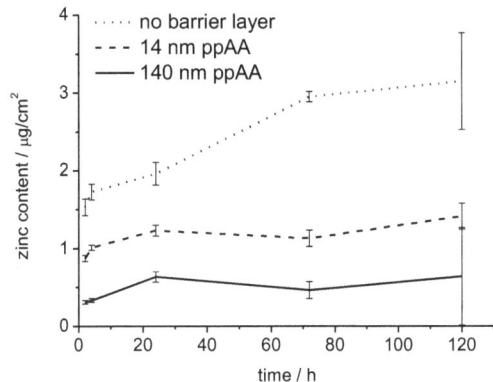

Figure 5-4. Concentration of zinc in the leachate from bilayers of ppAA on ppZn(acac)2 over the course of 120 h.

5.3 Antimicrobial Efficacy of Bilayers with ppAA on ppZn(acac)$_2$

According to what is known from the literature, concerning zinc the minimum inhibitory concentrations (MIC) for bacteria lie in the µg/ml range.[177] The ppAA top-layer must not be too thick so that sufficient amounts of the antimicrobial agent are released. Therefore, ppAA films with 20 nm thickness were chosen for the following experiments.

5.3.1 Bacterial Colony Counting from Bilayers with ppAA on ppZn(acac)$_2$

Figure 5-5 shows the result of plating the supernatant of *E. coli* suspensions (with 10^6 CFU/ml). One ml of the suspension was incubated with uncoated and coated glass cover slips overnight in 1/10 LB medium. Subsequently 20 µl were taken from the supernatant and a dilution series was made with dilutions of 1/1, 1/2500, 1/5000, and 1/10000. These were then plated and grown on agar overnight.

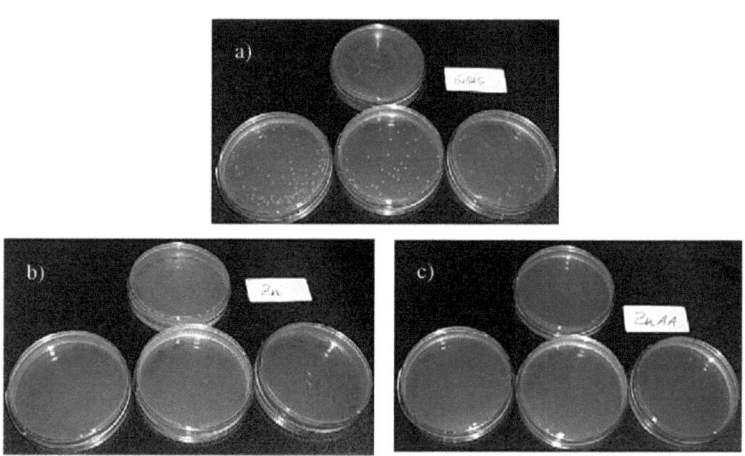

Figure 5-5. Plating of bacterial suspensions of E. coli after incubation overnight on a) uncoated and coated glass slides; b) ppZn(acac)2 and c) a bilayer of ppAA on ppZn(acac)2. A serial dilution was made from 20 □l with 1/1 (top), 1/2500, 1/5000, and 1/10000 (left to right).

Uncoated glass slides (a) did not inhibit bacterial growth. The plating of undiluted supernatant (petri dish on top) resulted in a dense lawn of bacteria. The glass slides coated with 500 nm of ppZn(acac)$_2$ as well as with the bilayer (500 nm of ppZn(acac)$_2$ and 20 nm of ppAA) showed no bacterial growth after overnight culture.

Also, *S. aureus* was incubated on untreated and treated polystyrene petri dishes. The procedure was similar to that used for *E. coli* except for the substrate (here: petri dishes) and the suspension volume which was 4 ml. The amount of colonies that grew on the agar plates were counted and are shown in Figure 5-6. Again, no growth of bacteria occurred after overnight culture with *S. aureus* on ppZn(acac)$_2$ as well as on the bilayer of ppZn(acac)$_2$ and ppAA.

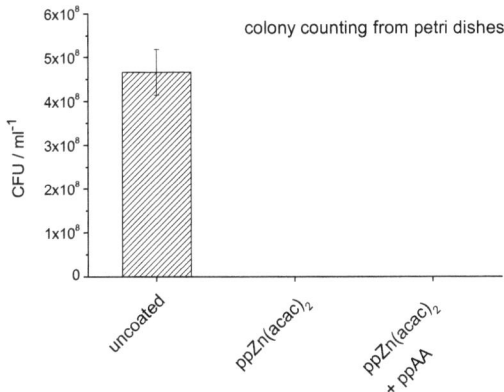

Figure 5-6. Colony counting of S. aureus after incubation overnight on uncoated and coated polystyrene petri dishes (ppZn(acac)2 and a bilayer of ppAA on ppZn(acac)2).

5.3.2 Coating Urethral Catheters with Bilayers of ppAA on ppZn(acac)$_2$

Bacterial testing was also performed on medical devices, like silicone urethral catheters (Coloplast, Denmark). The tubes were cut into 5 cm long pieces and were glued[10] vertically onto a sample holder with at least 1 cm distance to one another. Subsequently, the samples were coated with 500 nm of ppZn(acac)$_2$ or the bilayer. Due to the relatively small inner diameter of 3 mm it is unlikely that coating occurred on the inside of the tubes. However, this was not investigated except for visual inspection of the samples which revealed no change in color. This implied that the inner part of the catheters was not coated. Antimicrobial properties therefore entirely relied on the outer coating of the catheters. Samples were immersed in 4 ml of 1/10 LB medium containing 10^6 CFU/ml *S. aureus* in a 10 ml falcon tube. The samples were incubated overnight and the supernatant subsequently plated as described above.

---

[10] Epoxy glue (two-component, solvent free) was used so that the vacuum during plasma deposition was not disturbed.

Figure 5-7 shows the results of colony counting on the agar plates. Uncoated silicone catheters did not inhibit bacterial growth whereas coated samples for both tested layer systems did and no bacteria were found growing on the latter. The amount of zinc released from the coated part of the sample was not measured but was sufficient to cause bacterial death although the inner portion of the catheters remained uncoated. This is one of the advantages of an active release system compared to contact killing strategies. Antimicrobial properties are not locally restricted but are exhibited in the general area of the coated device. However, the release will also occur even without bacteria being present and the material released will be transported into surrounding parts of the tissue and body. Depending on the medical application this might be acceptable.

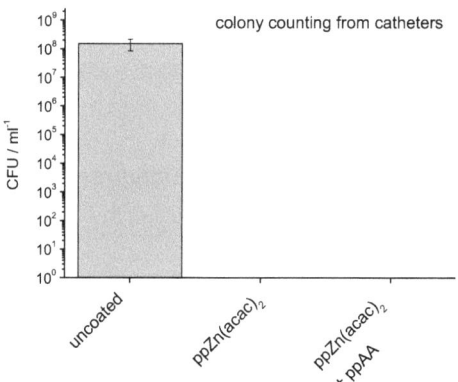

Figure 5-7. Colony counting of S. aureus incubated in a falcon tube together with uncoated and coated silicone catheter samples.

5.3.3 Cell Adhesion on Bilayers with ppAA on ppZn(acac)$_2$

Antimicrobial properties are desired for many biomedical applications but antimicrobial agents may also be cytotoxic. This is also the case for zinc and the concentration of this antimicrobial metal has to be balanced against the potential cytotoxicity. Figure 5-8 shows calcein stained HUVECs on uncoated and coated titanium discs (99.7 alloy, Alfa Aesar, Karlsruhe, Germany). The procedures for the cell experiments were the same as described in section 3.1.4. A 10 nm ppHMDSO layer was deposited on titanium prior to further surface modifications in order to enhance adhesion and stability of the subsequent layers as described in section 3.1.1. The microscopic images show that endothelial cells easily attached to bare titanium and that cell attachment was enhanced by coating the samples with ppAA. This is indicated by the well spread morphology of the cells.

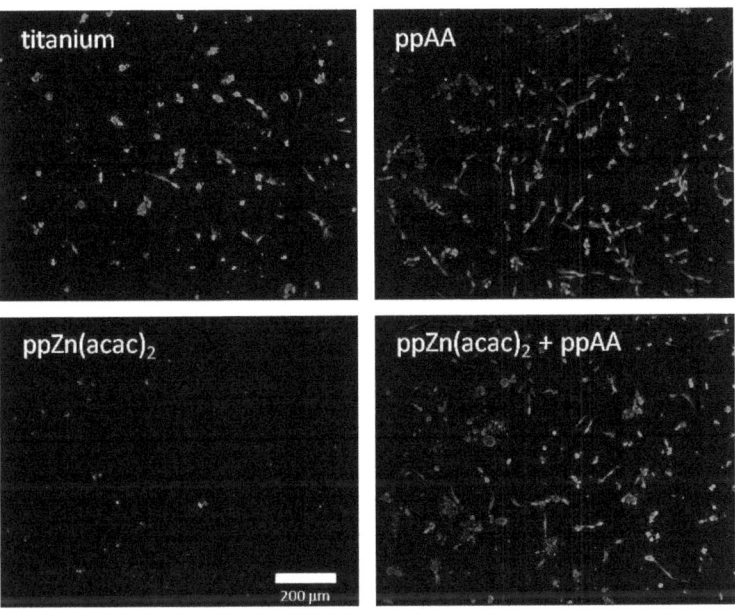

Figure 5-8. HUVECs on uncoated and coated titanium samples after 24 h.

Depositing a 500 nm thick layer of ppZn(acac)$_2$ on titanium reduced cell adhesion. Cell numbers were low and the few cells present have spherical shapes and do not indicate healthy or propagating cells. Whether this is an apoptotic or necrotic effect was not investigated here.

The plasma polymer modification with Zn(acac)$_2$ can be altered by deposition of a 20 nm ppAA film thus obtaining a bilayer. Cell adhesion was improved for such a bilayer compared to ppZn(acac)$_2$ and was comparable to that of ppAA only.

However, in contrast to ppAA only, the bilayer not only exhibited cell-adhesive properties but at the same time showed antimicrobial efficacy as well (chapter 5.3). Therefore, a combination of both antimicrobial and cell-adhesive properties can be obtained by adjusting the thickness of ppZn(acac)$_2$ and ppAA plasma polymer films. However, so far the antimicrobial effect is restricted to gram-positive bacteria. Chapter 6 will address this issue.

The images were also analyzed with the ImageJ software as described in section 3.1.4. Figure 5-9 shows the cell count from HUVECs on uncoated and coated titanium discs. The results are in good agreement to what has been already described. The cell number more than doubled as a result of coating titanium with ppAA. For 500 nm of ppZn(acac)$_2$ the amount of cells drastically decreased whereas the bilayer of the zinc containing film and ppAA increased the cell number to that found for bare titanium.

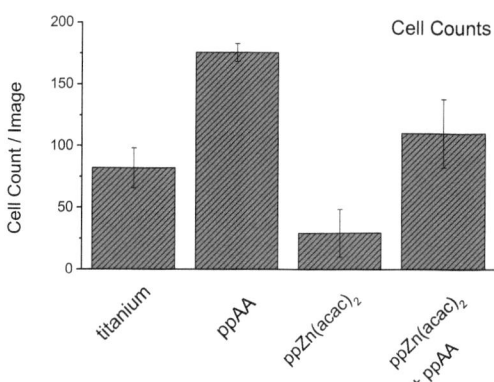

Figure 5-9. Cell count evaluated from images shown in Figure 5-8.

Looking at the surface coverage confirms these results (Figure 5-10). The total surface covered by cells was more than three times as large comparing ppAA with bare titanium. This is accounted to the morphology of the cells since the cell numbers only doubled. Surface coverage of cells on 500 nm ppZn(acac)$_2$ was almost zero whereas the bilayer of ppAA on ppZn(acac)$_2$ again increased cell coverage to a level double of that seen for bare titanium. This also was caused by the cell morphology since the cell count was about the same. Considering the results, the presented bilayer system of ppZn(acac)$_2$ and ppAA is a feasible antimicrobial and cell-adhesive coating for devices in medical applications.

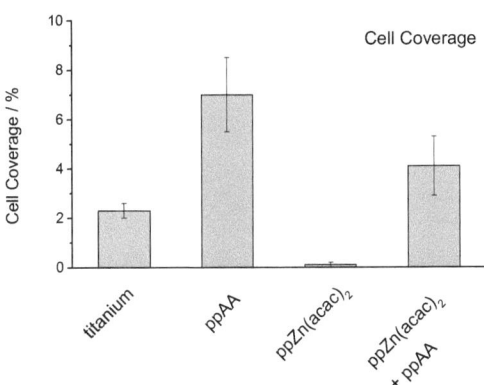

Figure 5-10. Cell coverage evaluated from pictures shown in Figure 5-8.

5.3.4 Cell Adhesion on Biofunctionalized Bilayers with ppAA on ppZn(acac)$_2$

So far in this work no cell behavior on fibronectin modified bilayers has been described and other cell types were not considered yet. A very common cell type used for cell experiments is the mouse cell line *fibroblast NIH 3T3*. These cells were seeded on uncoated and coated glass. The glass cover slips had a diameter of 13 mm and were put into 24-well plates. In this case 10,000 fibroblasts were incubated overnight on the samples in 1 ml of Dulbecco's Modified Eagle Medium (DMEM) with 10% FCS. Figure 5-11 shows micrographs of the cells after 24 h. The 500 nm of ppZn(acac)$_2$ did not allow for adhesion of cells indicated by the spherical morphology compared to uncoated glass. In contrast to this, ppAA and ppAA on ppZn(acac)$_2$ caused good adhesion of cells comparable to that of the control. However, upon immobilizing fibronectin on the bilayer as described in section 3.1.3 cell adhesion was drastically increased and cell-cell interaction took place due to the high number of cells and the large area they covered. The cells formed almost a confluent layer on fibronectin coated bilayers of ppZn(acac)$_2$ and ppAA. It is therefore concluded that cell adhesion is possible on the bilayers but can be improved strongly by immobilization of fibronectin on such coatings.

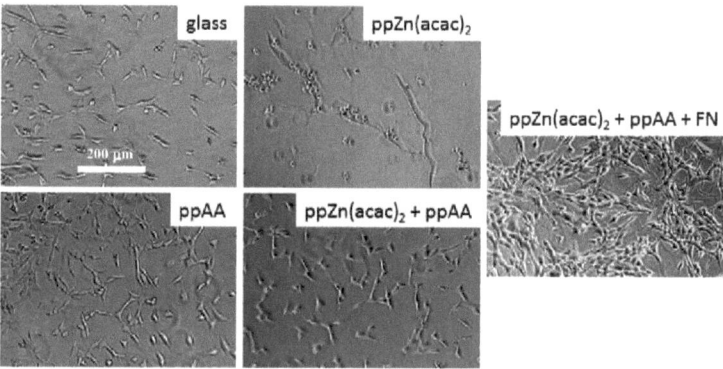

Figure 5-11. Fibroblasts NIH 3T3 on uncoated and coated glass cover slips after 24 h of incubation.

Due to the way the microscopic images were taken no automated evaluation with ImageJ could be conducted.[11] Nevertheless, cell numbers were counted manually and results are shown in Figure 5-12. The number of cells adherent to the ppAA surface was almost 3-fold higher compared to that on bare glass. The error was large which is not only due to the manual counting but can also be contributed to the fact that the glass cover slips are so light in weight that they never perfectly align with the bottom of the well (24-well plate). Cells in most cases accumulated in some portions of the well and were not homogeneously distributed.

Nevertheless, the 500 nm films of ppZn(acac)$_2$ considerably prevented cell adhesion and spreading whereas the bilayer exhibited cell adhesion and spreading comparable to that of uncoated glass. Immobilization of fibronectin on the bilayer resulted in cell numbers comparable to that on ppAA. The effect of fibronectin on the cell count was not as pronounced as compared to endothelial cells (section 3.1.4).

---

[11] Automated procedures rely on high contrast images that enable the algorithm to distinguish between particles and background.

This might be due to the fact that for the experiments with HUVECs fibronectin was immobilized on ppAA only without any ppZn(acac)$_2$ coating. Nevertheless, cell morphology on the fibronectin modified bilayers implied healthy cells and they were almost exhibiting a confluent layer. It is possible that the zinc released into the medium reduced cell proliferation without compromising the strong adhesion promoting effect of fibronectin. Whether this is true could be investigated by using proliferation assays which were not conducted here.

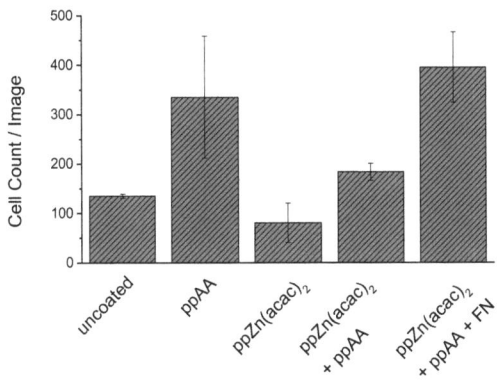

Figure 5-12. Cell count evaluated from images shown in Figure 5-11.

6 Multilayered Coatings with silver, ppZn(acac)$_2$, and ppAA

As described in section 4.2.1 the antimicrobial efficacy of ppZn(acac)$_2$ strongly depends on the bacterial species used in the experiment. Zinc containing films can therefore not be the only solution to prevent microbial colonization. The aim of this chapter is to broaden the spectrum of antimicrobial efficacy by an additional layer of silver below the already mentioned bilayer. The coating would thus contain a multilayer of up to six modifications. Fortunately, most of the modification procedure takes place under the same vacuum conditions so that, with the right equipment, expenses in industrial application would be minimized.

The substrate was coated with thermally evaporated silver which is known for its antimicrobial properties.[181] In order to obtain better adhesion of plasma polymer deposits the silver coated samples were then immersed in an ethanolic solution of 50 mM allyl mercaptane as described in section 3.1.3. Subsequently 500 nm of ppZn(acac)$_2$ were deposited followed by 20 nm of ppAA. The ppAA was also wet chemically modified with DEGDGE in order to immobilize fibronectin. A basic scheme of the layer system discussed in this chapter can be seen in Figure 6-1.

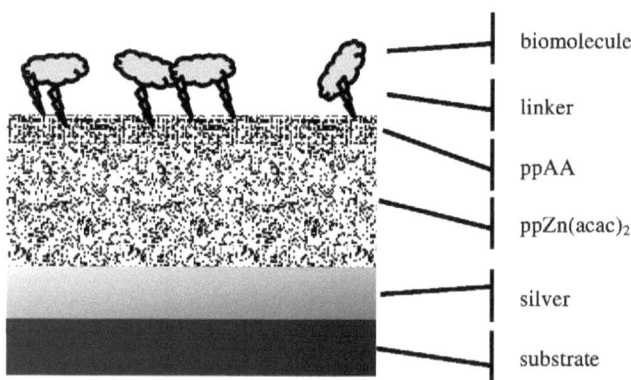

Figure 6-1. Scheme of the layer system discussed in chapter 6.

6.1 Film Analysis of Multilayered Coatings with silver, ppZn(acac)$_2$, and ppAA

In principle a thicker layer of thermally evaporated silver would not lead to a gain of silver ions released since the supernatant only comes in contact with the surface of the silver (2-dimensional system). This is in contrast to ppZn(acac)$_2$ where the aqueous solution diffuses into the porous film forming a 3-dimensional system. Nevertheless, the silver coating needs to be sufficiently thick to provide homogeneous coverage.

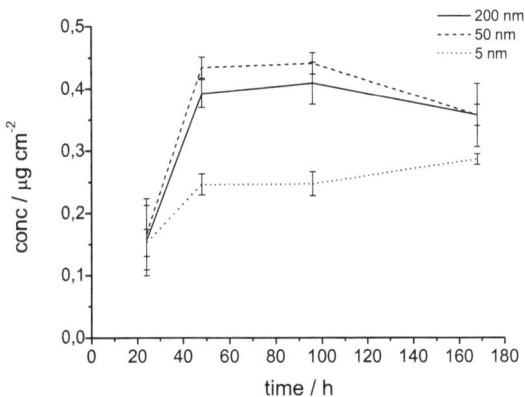

Figure 6-2. Silver content in the supernatant on thermally evaporated silver with varying thickness.

Figure 6-2 confirms this by showing the silver content in the supernatant of silver coatings depending on the thickness of the silver film. Ultrapure water was used for these experiments in order to circumvent any effects caused by other ions in the solution. The supernatant was collected at each point in time and the sample reincubated with ultrapure water. The silver was deposited on round glass cover slips with a diameter of 13 mm. After 2 d the release of silver from the surface remained constant. Also, 5 nm thin silver films did not seem to provide a complete coverage of the substrates since the silver release never reached the extent seen for the thicker films. Nevertheless, a 50 nm silver layer was sufficient for a constant release that could not be increased by thicker layers of silver.

Figure 6-3 shows an SEM image of a multilayer cross section on a Si-wafer. The wafer was first coated and subsequently broken in half. The breaking edge was then observed in the microscope (normal to the breaking edge). The black arrow points at the silver film positioned above the polished Si-wafer. The film on top consisted of the 500 nm ppZn(acac)$_2$ as well as the 20 nm of ppAA. Nevertheless, SEM did not resolve the ppAA layer on ppZn(acac)$_2$. This is mainly due to the lack of contrast between the two films. It is possible that with increasing zinc content in ppZn(acac)$_2$ (e.g. higher input power upon deposition[95]) this contrast increases as well and distinction would be possible in SEM imaging. However, this was not investigated.

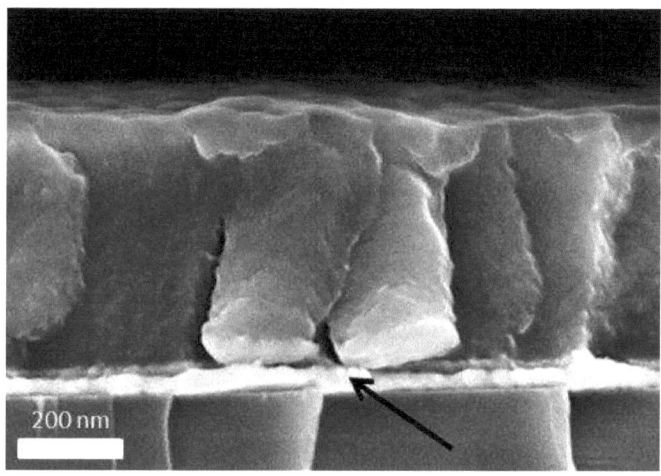

Figure 6-3. SEM cross section of a Si-wafer coated with 50 nm of silver (black arrow), 500 nm of ppZn(acac)2, and 20 nm ppAA. The barrier layer of ppAA was not resolved possibly due to lack of contrast.

Additionally to SEM, the same samples were analyzed with EDX in order to obtain further information about the composition of the multilayers. Figure 6-4 shows the elements that were found in the multilayer.[12] The interface of the Si-wafer and the silver film was set to zero (see x-axis). At this point the signal for silicon (grey) decreased whereas that of silver (blue) increased. Other elements were observed as well possibly originating from contaminations. However, scattering of electrons reduces the resolution and lead to overlapping and broadened signals. Nevertheless, a clear distinction between silver (blue) and zinc (green) can be made at the interface at a distance of 50 nm from the surface of the Si-wafer. Oxygen as well as carbon signals (red and black) increased in the area of the plasma polymer deposit. It is unknown why the carbon signal (black) is so strongly increased towards 500 nm representing the top of the multilayer. Nevertheless, the zinc content remains constant up to 500 nm. At this point the signals for zinc (green) and oxygen (red) decreased. Due to the little amount of ppAA compared to the rest of the multilayer no sufficiently distinct signal for nitrogen was detected (except for background) and is therefore not shown here.

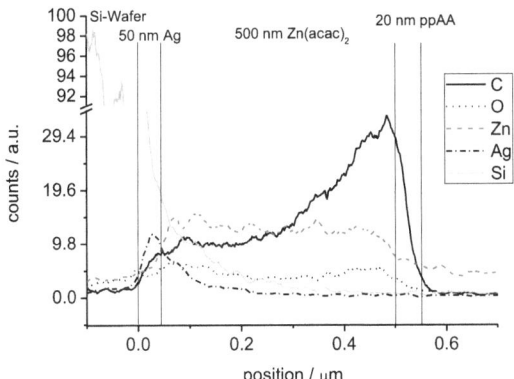

Figure 6-4. EDX cross section of the sample shown in Figure 6-3. Nitrogen is not shown because of the amount of background signal observed for nitrogen.

---

[12] This is true except for hydrogen and nitrogen. Whereas the former cannot be detected the latter gave high amounts of background. It is worth noting here that nitrogen can be found as contamination in most samples and the resolution of EDX was not sufficient to resolve the 20 nm thin ppAA film.

In addition, film analysis included XPS which has a sampling depth of approximately 10 nm and therefore is suitable for the investigation of the surface of a specimen. All layer systems were analyzed concerning their elemental composition at the surface and the data is shown in Figure 6-5. The original data can be found in Table 6-1. Silver films showed large amounts of carbon which is mainly due to impurities. Freshly prepared thermally evaporated silver films readily adsorb airborne contaminants and can also oxidize partially. Due to the fact that XPS is very surface sensitive these impurities contribute to the signal to a large extent.

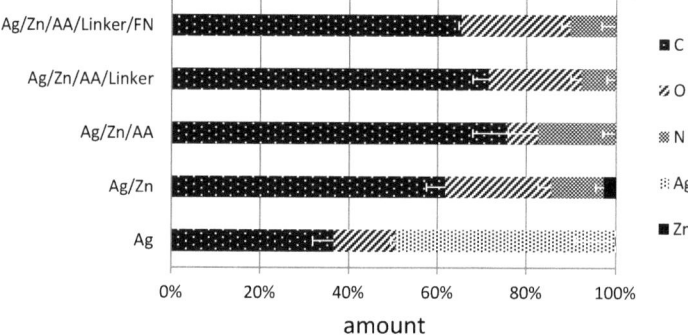

Figure 6-5. XPS survey spectra of the single layer systems showing elemental composition (normalized to 100%) in atomic percent. Error bars for each element extend into the corresponding column for that element.

Table 6-1. The original data set for Figure 6-5 with elemental composition in percent without normalization (difference to 100% is attributed to impurities, e.g. Na, Cl, Si)

surface	at % C	at % O	at % N	at % Ag	at %Zn
Ag/Zn/AA/Linker/FN	61.3 ± 0.6	22.8 ± 1.8	9.6 ± 3.0	-	-
Ag/Zn/AA/Linker	66.9 ± 3.4	19.3 ± 2.2	7.3 ± 1.9	-	-
Ag/Zn/AA	75.2 ± 7.8	6.8 ± 6.7	17.2 ± 2.9	-	-
Ag/Zn	60.8 ± 4.2	23.5 ± 2.9	11.5 ± 1.8	-	2.7 ± 0.6
Ag	33.3 ± 4.3	12.9 ± 1.0	-	45.0 ± 8.3	-

After the deposition of 500 nm of a ppZn(acac)$_2$ layer surface composition changed and no silver was detected any more. The carbon as well as the oxygen content increased, whereas zinc was present only in small amounts. This corresponds well to the literature where similar films were investigated.[95] Looking at the elemental composition of the Zn(acac)$_2$ monomer which consists of carbon, oxygen and zinc in a ratio of 60:20:6 the composition of the plasma deposit agreed reasonably well with 60:25:3. Nitrogen was also found but was probably a result of contamination from residual air during plasma polymerization. Also, the precursor was used without further purification and might be a source of volatile nitrogen compounds. Further investigations should address this issue but these impurities did not impede antimicrobial properties (see section 4.2). Also, the amount of zinc in the deposit was 50% lower compared to the precursor. Higher amounts of zinc can be reached by using higher input powers.[95] However, this also leads to higher zinc release rates and would make control over the release increasingly difficult.

Deposition of 20 nm of ppAA on top of the above mentioned films resulted in spectra that did not show any zinc which is expected due to the low sampling depth of XPS (about 10 nm). In accordance with the precursor composition both the carbon and nitrogen content were higher compared to ppZn(acac)$_2$. The precursor composition is 75:25 for carbon and nitrogen and 75:17 was found for the ppAA coated samples in XPS. Airborne contaminants as well as gaseous oxygen may have reacted with reactive species on the film. The results of quantifications of primary amino functional groups were already discussed in section 3.1.2.

After wet chemical treatment with DEGDGE the nitrogen content decreased as the linker does not contain any nitrogen. However, due to the short length of the linker ppAA may contribute to the nitrogen signal. The sampling depth of XPS was sufficient to resolve parts of the ppAA even after coupling of the linker. The amount of oxygen was increased. The carbon to oxygen ratio in DEGDGE is 66:33 which corresponds reasonably well with 70:25 considering the fact that ppAA contributed to the signal thus increasing the carbon signal.

Immobilization of fibronectin did not show signals sufficiently distinct from that of DEGDGE on ppAA. The same is true for control samples that were incubated with fibronectin without using the DEGDGE linker (not shown here). It is possible that the surfaces are not homogeneously covered by fibronectin and that agglomeration of protein may play a role. Nevertheless, this did not hamper cell adhesion (see section 3.1.4).

6.2 Release of Silver and Zinc from Multilayered Coatings with silver, ppZn(acac)$_2$, and ppAA

Silver exhibits antimicrobial properties[181] but is also known for its cytotoxic effects[182]. It is therefore important to control the release of silver from the surface. The principle applied here is similar to the approach shown for the control of the release of zinc. Here, not only the ppAA film functions as a barrier layer but the ppZn(acac)$_2$ layer as well. Both are positioned on top of the silver film and therefore can limit the amount of silver released from the multilayer.

Figure 6-6 shows the amount of zinc released from the multilayers with three different layers of ppZn(acac)$_2$ of varying thickness. For each point in time the supernatant was collected and the sample incubated again with ultrapure water. For all samples a strong decrease of zinc release within two days was observed. After this time the release of zinc from the thinner films (100 and 250 nm) was almost zero, whereas that of the 500 nm thick films slowly decreased from 0.75 to 0.4 $\mu g/cm^2$ over the course of three weeks. There is no clear distinction between the behavior of the 100 and the 250 nm films. It is likely that the thickness measured did not match the thickness on the samples used for the release experiments. This is possible because thickness measurements were conducted on Si-wafers whereas samples for release measurements were glass slides. These samples were placed close but not at the exact positions during the deposition. Looking at section 4.1.1 shows that minor differences in position can lead to strong deviations in film thickness. This is especially true for thinner films where control of film thickness depends on exact timing of the plasma polymerization process. This problem could be overcome by using smaller samples so that all the samples can be placed in a zone of same deposition rate. However, this would lead to smaller sample volumes insufficient for ICP-OES measurements.

Nevertheless, the results show that the principle of controlling the release of zinc from the surface depended on the thickness of ppZn(acac)$_2$. Due to its ability to absorb water, which in turn dissolves the zinc, thicker films provide a greater reservoir for metal release. Over time this release was reduced due to the permanent loss of zinc from the film. Since thicker films represent a greater reservoir for zinc the release of material is prolonged.

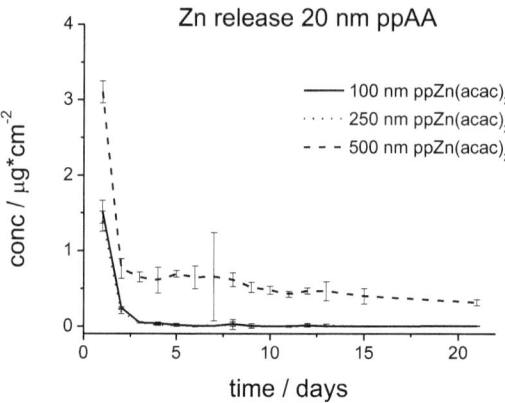

Figure 6-6. Zinc content in the leachate of a multilayer consisting of 50 nm of silver, ppZn(acac)2 of varying thickness, and 20 nm ppAA.

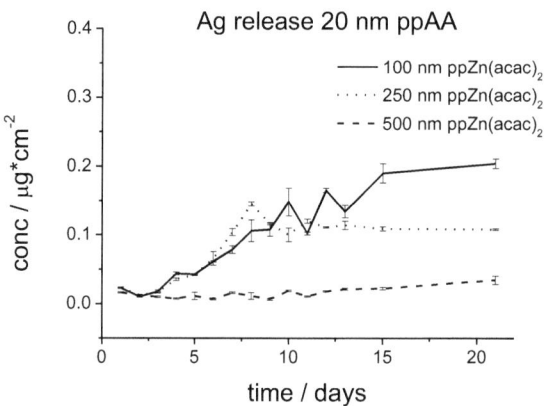

Figure 6-7. Silver content in the leachate of a multilayer consisting of 50 nm of silver, ppZn(acac)2 of varying thickness, and 20 nm ppAA.

The silver release was also measured for the same multilayers (Figure 6-7). Depending on the thickness of ppZn(acac)$_2$ films the release of silver was retarded. Thicker ppZn(acac)$_2$ films exhibited slower release rates for silver because the plasma deposit acts as a barrier layer. Over time the release was increased for all samples since the aqueous solution diffused through the zinc containing plasma polymer films and reached the silver film. Permeability of the layers was increased with decreasing zinc content.

The results shown in this section indicate that the release of metals from a multilayered coating as it is presented in this work can be controlled by various parameters. The thickness of the zinc containing plasma polymer as well as the barrier layer of ppAA can influence this release. This way, by controlling the thickness of these layers, the antimicrobial and cell-adhesive properties of such a multilayer can be controlled.

How the release of a multilayered coating is influenced by immobilization of fibronectin was not investigated. However, wet chemical treatment of these layers lead to a release of metal and therefore has to be kept as short as possible. Antimicrobial efficacy and longevity of the presented multilayers (including fibronectin coated multilayers) will be discussed in the next section.

6.3 Antimicrobial Efficacy of Multilayered Coatings with silver, ppZn(acac)$_2$, and ppAA

Multilayered coatings were tested for antimicrobial properties using *E. coli*, *S. aureus*, and *P. aeruginosa*. Various combinations of multilayer compositions were investigated. Silver films had a thickness of 50 nm, ppZn(acac)$_2$ films were 500 nm and ppAA barrier layers 20 nm in thickness.

6.3.1 Bacterial Testing for *E. coli* on Multilayered Coatings with silver, ppZn(acac)$_2$, and ppAA

A simple test for antimicrobial efficacy was utilized in the same manner as described in section 5.3.1 where *E. coli* was incubated on uncoated and coated glass cover slips overnight. The supernatant was then transferred on agar in two different dilutions. Figure 6-8 shows photographs of the agar plates and clearly reveals the antimicrobial properties of the multilayered coating containing silver, ppZn(acac)$_2$, and ppAA (c) compared to the controls (polystyrene petri dishes (a) and uncoated glass slides (b)). Furthermore, immobilization of fibronectin on the multilayer (d) did not reduce the antimicrobial effect at least after 24 h.

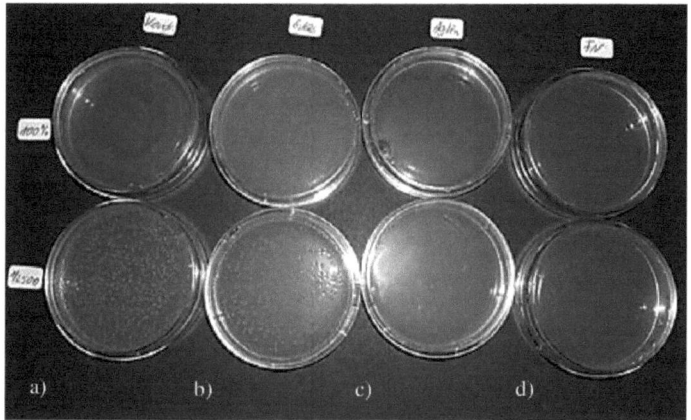

Figure 6-8. Agar plates after incubation with supernatant containing E. coli. The supernatant was previously incubated on a) uncoated petri dishes, b) uncoated glass slides, c) Ag/ppZn(acac)2, and d) Ag/ppZn(acac)2/ppAA/FN coated glass slides. The two rows correspond to two dilutions of the supernatant.

6.3.2 Coating Non-Woven Fabrics with Multilayered Coatings with silver, ppZn(acac)$_2$, and ppAA

The application of the multilayer coating on a clinically relevant material, used for wound dressings, was also conducted. Non-woven PET from Freudenberg (Weinheim, Germany) was coated with the multilayer system. Combinations of silver, ppZn(acac)$_2$, ppAA, and immobilized fibronectin were used.[13] Figure 6-9 shows photographs of the uncoated (a) and coated (b) fabric. The greyish color originated from the silver coating (this is also seen for commercially available plasters and band aids that contain silver) whereas the brownish tint resulted from additional layers of ppZn(acac)$_2$ and ppAA. It is worth noting that the material remained as flexible and soft as the uncoated control. This is due to the fact that the plasma deposition procedure does not influence the bulk properties of the substrate.

---

[13] Whether a wound dressing should bear such cell-adhesive moieties was not studied here.

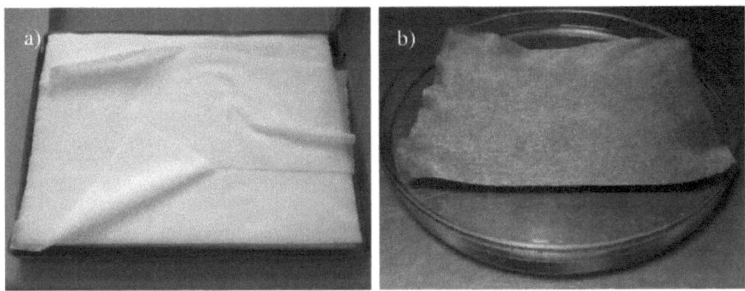

Figure 6-9. Photographs of a) uncoated and b) multilayer coated non-woven PET. The multilayer consisted of silver + ppZn(acac)2 + ppAA + linker bound fibronectin.

EDX spectra (Figure 6-10) showed that only carbon and oxygen were detected on uncoated PET, whereas additional signals appeared on the multilayered coatings. Silver as well as zinc were present in the sample; also minor amounts of nitrogen and sulfur were found. The latter originates from the interstitial layer of allyl mercaptane between silver and ppZn(acac)$_2$ (used for adhesion improvement as described in section 2.2). Furthermore, sulfur is also present in proteins like fibronectin. Nitrogen originated from ppAA as well as from fibronectin.

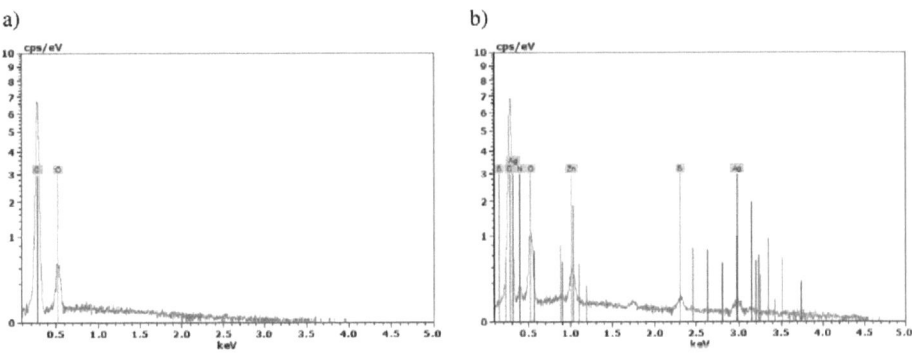

Figure 6-10. EDX spectra of the woven PET samples shown in Figure 6-9.

Both the untreated and treated PET samples were incubated overnight with 4 ml of a suspension containing 10^6 CFU/ml of *S. aureus* in 1/10 LB medium in a polystyrene petri dish. Subsequently, the samples were stained with Baclight Live/Dead (Invitrogen, Germany) and optical micrographs were taken with the appropriate filters (Figure 6-11). As already seen for samples in section 4.2.2, background staining was also visible here. However, stained spots and particles represent bacteria. Whereas the uncoated sample (a) showed green staining on the fibers, the multilayer coated material exhibited stained bacteria preferentially in red. With regard to *S. aureus* the multilayer clearly acts antimicrobial over the course of at least 24 h.

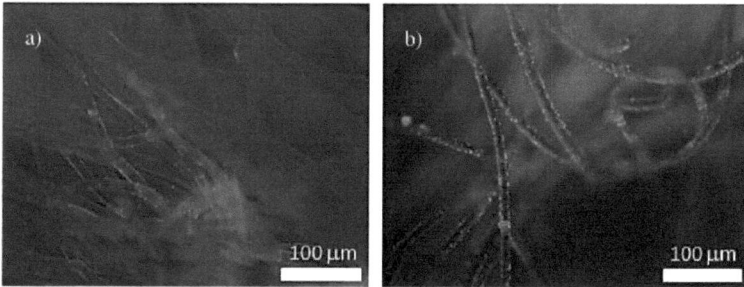

Figure 6-11. Live/Dead staining of S. aureus on a) uncoated and b) coated woven PET shown in Figure 6-9.

6.3.3 Longevity of Antimicrobial Efficacy of Multilayered Coatings with silver, ppZn(acac)$_2$, and ppAA

So far antimicrobial efficacy was tested in overnight cultures, which is not sufficiently long to judge the feasibility of the antimicrobial surface modifications for medical applications. It is therefore necessary to test the longevity of these properties. Uncoated and coated PTFE discs with a diameter of 13 mm were incubated overnight with 10 µl of a suspension containing 10^6 CFU/ml bacteria in 1/10 LB medium. The complete suspension was then collected, plated on agar, and incubated overnight. The sample was again incubated with a fresh bacterial suspension on the same spot. The agar plates were visually inspected for bacterial growth after the overnight culture. The number of days with no bacterial growth was observed.

Figure 6-12 shows the results for *S. aureus*. The controls that do neither contain silver nor zinc did not inhibit the growth of bacteria at all. Coatings from silver, zinc as well as the combination of both lost their antimicrobial properties after about one week. After that time the release of metal was too low to cause bacterial death. The addition of a ppAA barrier layer on top of these surfaces increased the longevity of antimicrobial properties. After two weeks the metal content seemed to be exhausted and not sufficient to cause bacterial death. The bilayer made from ppZn(acac)$_2$/ppAA/fibronectin, however, showed antimicrobial activity for only one week. This is due to the fact that wet chemical treatment (immobilization of fibronectin) involves 1 h of exposure to aqueous solutions and additional washing steps. This procedure already causes a loss of antimicrobial agent, most of which occurs in the first minutes to hours (compare Figure 6-6 in chapter 6.2). This is also true for the samples coated with silver/ppZn(acac)$_2$/ppAA/fibronectin but the additional release of silver probably compensated for the initial loss of zinc.

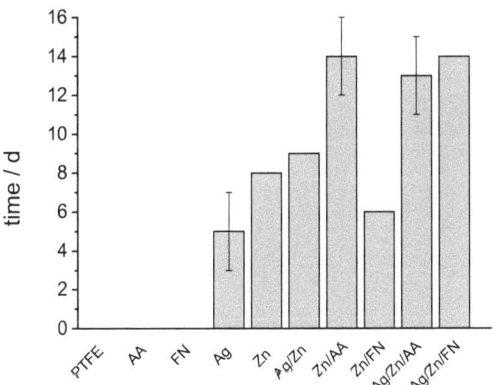

Figure 6-12. Qualitative evaluation of the longevity of the antimicrobial efficacy for various coatings on PTFE on S. aureus.

As already explained in section 4.2.1 (compare Figure 4-4) the zinc-based antimicrobial coatings are not effective against gram-negative bacteria like *P. aeruginosa*. Figure 6-13 confirms this but at the same time reveals the applicability of multilayers containing silver. It is worth noting that the samples of a particular coating showed no variation in their longevity (no error bars). This is due to the fact that the samples were obtained from one batch during coating procedure and thus were probably very homogeneously coated. Also, the very roughly distributed measurements in time (days) contributed to this.

The experiment revealed that antimicrobial properties lasted no longer than 12 d. As described before *P. aeruginosa* is tolerable towards many metals and may therefore colonize the samples quicker compared to gram-positive bacteria.[183,184] Multilayers containing a top-layer of ppAA did not increase the longevity of the samples since the release of metal was decreased too far to exhibit antimicrobial activity against *P. aeruginosa*. At the same time, the metal content of multilayers without barrier layer was depleted relatively fast. It is therefore concluded that multilayers containing a barrier layer still need a lot of experimentation in order to optimize the thickness of the various layers. Gram-negative bacteria seem to be especially challenging with respect to the exact metal release behavior of the coating, especially when cytotoxicity has to be considered as well.

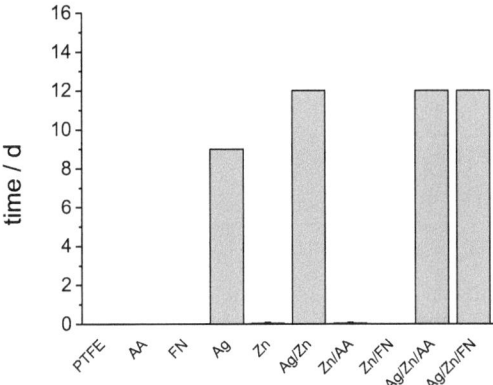

Figure 6-13. Qualitative evaluation of the longevity of the antimicrobial efficacy for various coatings on PTFE on P. aeruginosa.

7 Conclusion and Outlook

Surface modifications for medical applications play an important role in modern medicine. Medical devices are fabricated from a large variety of materials, e.g. titanium, PET, or PTFE. Biointegration is one of the big issues in implantology. Integration of artificial materials is often hampered by foreign body reactions triggered by the reaction of surrounding tissue with implant failure being the result. Wound healing is further impaired by the fact that cavities between tissue and material offer pathways for bacteria to enter the wound. Nosocomial infections with fatal outcome are often seen in hospital scenarios. Coatings on medical devices and implants that exhibit cell-adhesive as well as antimicrobial properties are therefore desired. Offering a tool to overcome repulsion of medical devices as well as the inclusion of antimicrobial properties was the aim of this work.

Plasma polymerization offers a way for the surface modification of various materials. This technique is suitable for coating prefabricated devices in medical applications. The method was utilized in this work in order to combine antimicrobial and cell-adhesive surface modifications aimed at minimization of foreign body reactions as well as reduction of hospital acquired bacterial infections.

Plasma polymer films were analyzed with respect to their static water contact angles because biological systems require aqueous solutions and the release of metal from antimicrobial layers is based on dissolution of the metal. A method to enhance cell-adhesive properties was presented by the wet chemical modification of ppAA films. Fibronectin, a protein of the extracellular matrix was immobilized to specifically target biochemical pathways for cell-surface interactions. Infrared absorption techniques were applied in order to characterize the plasma polymer deposits and wet chemical modifications. In addition, kinetic SPR measurements helped to observe the chemisorption of fibronectin.

The cell-adhesive film also served as a tool to control antimicrobial properties of the surface since it reduced the release of metal from the sub-layers. SEM and EDX were utilized to characterize the multilayered system. Furthermore, XPS provided information about the elemental composition of the multilayer and was also used to investigate the amount of primary amino functional groups on ppAA. Sufficient amounts of primary amines were necessary for the immobilization of fibronectin on the plasma polymer deposit.

Because antimicrobial efficacy of the presented coatings was based on the release of metal from the surface ICP-OES was used to determine silver and zinc concentrations in the films as well as the release from the surface upon immersion in water. It was shown that the release can be controlled by varying film thickness of both the metal containing layer as well as of the barrier layer on top.

Cell-adhesive properties and cytotoxicity were tested by using fibroblasts and endothelial cell lines. Their morphology, number, and surface coverage were observed with optical microscopy and quantified. It was shown that immobilization of fibronectin strongly enhances cell adhesion. Nevertheless, thick layers of ppZn(acac)$_2$ exhibited cytotoxic effects. This was circumvented by applying a barrier layer on top that reduced the release of the metal from the surface.

Antimicrobial surface modifications based on zinc were shown to be effective against *E. coli* and *S. aureus* but not against *P. aeruginosa*. The latter is known for its resistance against zinc and a further metal containing layer was added by depositing a silver film on the substrates prior to coating with ppZn(acac)$_2$ and ppAA.

The controlled release of metal from the coating served as a tool to reduce cytotoxicity. This offered the opportunity to deposit thicker films containing a large reservoir of antimicrobial agent. Thus, the longevity of antimicrobial efficacy of such deposits was greatly enhanced without compromising cytocompatibility.

Future work should be aimed at stability of the coatings especially towards handling in surgery scenarios. High shear forces along the coating can play a role for dental implants or vascular grafts. For example, delamination of the coating from a vascular graft could lead to an immediate thrombosis with acute risk of fatal outcome. Surface modifications have to withstand mechanical stress and different solutions and suspensions with varying pH may come into contact with the implant material. Furthermore, standardized sterilizing procedures (e.g. autoclaving, UV irradiation, ethanolic solutions) can cause further stress to the coating increasing the risk of losing the designed properties.

Experiments on cell adhesion and cytotoxicity, especially for the silver containing multilayer designs, need to be conducted. This may include viability and proliferation assays to quantify the cytotoxicity. Approved standards for testing procedures would have to be considered before such a coating can be utilized in medical applications. It is furthermore interesting to test the ppZn(acac)$_2$ and multilayer deposits for the induction of apoptosis and necrosis. So far it is unknown what the major contribution to cell death is. Additionally, cell cultures with immune cells or co-cultures with various cell types may offer new information on biocompatibility. It would also help to understand and prevent implant failure for devices coated with the proposed surface modifications. Additional types of cells depending on the application aimed for need to be considered, e.g. osteoblasts are especially interesting as a cell type to test for biointegration of dental implants.

There is far more bacteria responsible for hospital acquired infections than the ones discussed in this work. It would therefore be interesting to investigate the behavior of various types of bacteria on the presented surface modifications. For example, *S. aureus MSSA 476* (methicillin sensitive S. aureus) was used in this work, but *MRSA* (methicillin resistant S. aureus) strains are considered far more resistant to known antibiotics and would therefore be an ideal species to test for. Nevertheless, such strains require higher security standards for laboratories. Also co-cultures of various types and strains of bacteria, also with mammalian cell lines, could offer more information on combining antimicrobial and cell-adhesive properties.

Furthermore, wet chemical methods may offer a route to immobilize quaternary ammonium compounds as well as PEG-like brush polymers on the ppAA surface. By co-immobilizing various compounds new coatings with tailored properties could be designed. Also, the combination of different antimicrobial approaches, e.g. release systems as well as anti-fouling properties could increase antimicrobial efficacy. Multiple strategies would target various sites in the bacteria thus increasing bactericidal properties.

References

1   Hench, L. L. Biomaterials: a forecast for the future. *Biomaterials* **19**, 1419-1423 (1998).

2   Park, J. B. & Lakes, R. S. *Biomaterials: an introduction*. (Springer Verlag, 2007).

3   Mayhall, C. G. The epidemiology of burn wound infections: Then and now. *Clinical Infectious Diseases* **37**, 543-550 (2003).

4   Kunln, C. M., Douthitt, S., Dancing, J., Anderson, J. & Moeschberger, M. The association between the use of urinary catheters and morbidity and mortality among elderly patients in nursing homes. *American journal of epidemiology* **135**, 291 (1992).

5   Quirynen, M., Bollen, C., Papaioannou, W., Van Eldere, J. & van Steenberghe, D. The influence of titanium abutment surface roughness on plaque accumulation and gingivitis: short-term observations. *The International journal of oral & maxillofacial implants* **11**, 169 (1996).

6   Phaneuf, M. D. *et al.* Development of an infection-resistant, bioactive wound dressing surface. *Journal of Biomedical Materials Research Part A* **74**, 666-676 (2005).

7   Madigan, M., Martinko, J. & Parker, J. Brock Biology of Microorganisms (2000). *ISBN: 0-13-081922-0* (2000).

8   Heuer, O., Magiorakos, A., Gunell, M., Economopoulou, A., Blomquist, P. B., Brown, D., Walton, & C., P., N., Monnet, D. Antimicrobial resistance surveillance in Europe 2010. Annual report of the European Antimicrobial Resistance Surveillance Network (EARS-Net). (European Centre for Disease Prevention and Control, 2010).

9   Page, K., Wilson, M. & Parkin, I. P. Antimicrobial surfaces and their potential in reducing the role of the inanimate environment in the incidence of hospital-acquired infections. *J. Mater. Chem.* **19**, 3819-3831 (2009).

10 Bjarnsholt, T. *et al.* Why chronic wounds will not heal: a novel hypothesis. *Wound repair and regeneration* **16**, 2-10 (2008).

11 Roberts, R. R. *et al.* The use of economic modeling to determine the hospital costs associated with nosocomial infections. *Clinical Infectious Diseases* **36**, 1424 (2003).

12 Spellberg, B., Powers, J. H., Brass, E. P., Miller, L. G. & Edwards, J. E. Trends in antimicrobial drug development: implications for the future. *Clinical Infectious Diseases* **38**, 1279 (2004).

13 Maple, P., Hamilton-Miller, J. & Brumfitt, W. World-wide antibiotic resistance in methicillin-resistant Staphylococcus aureus. *The Lancet* **333**, 537-540 (1989).

14 Bouza, E., Garcia-Garrote, F., Cercenado, E., Marin, M. & Diaz, M. Pseudomonas aeruginosa: a survey of resistance in 136 hospitals in Spain. *Antimicrobial agents and chemotherapy* **43**, 981 (1999).

15 von Eiff, C., Becker, K., Machka, K., Stammer, H. & Peters, G. Nasal carriage as a source of Staphylococcus aureus bacteremia. *New England Journal of Medicine* **344**, 11-16 (2001).

16 Liu, G. Y. *et al.* Staphylococcus aureus golden pigment impairs neutrophil killing and promotes virulence through its antioxidant activity. *The Journal of experimental medicine* **202**, 209 (2005).

17 Cramton, S. E., Ulrich, M., Gotz, F. & During, G. Anaerobic conditions induce expression of polysaccharide intercellular adhesin in Staphylococcus aureus and Staphylococcus epidermidis. *Infection and immunity* **69**, 4079 (2001).

18 Cimolai, N. MRSA and the environment: implications for comprehensive control measures. *European Journal of Clinical Microbiology & Infectious Diseases* **27**, 481-493 (2008).

19 Klevens, R. M. *et al.* Invasive methicillin-resistant Staphylococcus aureus infections in the United States. *JAMA: the journal of the American Medical Association* **298**, 1763 (2007).

20 Labeda, D., Liu, K. C. & LE Casida, J. Colonization of soil by Arthrobacter and Pseudomonas under varying conditions of water and nutrient availability as studied by plate counts and transmission electron microscopy. *Applied and environmental microbiology* **31**, 551 (1976).

21 Yoon, S. S. *et al.* Pseudomonas aeruginosa Anaerobic Respiration in Biofilms:: Relationships to Cystic Fibrosis Pathogenesis. *Developmental cell* **3**, 593-603 (2002).

22 Livermore, D. M. Multiple mechanisms of antimicrobial resistance in Pseudomonas aeruginosa: our worst nightmare? *Clinical Infectious Diseases* **34**, 634 (2002).

23 Burdon, D. & Whitby, J. Contamination of hospital disinfectants with Pseudomonas species. *British medical journal* **2**, 153-155 (1967).

24 Favero, M., Carson, L., Bond, W. & Petersen, N. Pseudomonas aeruginosa: growth in distilled water from hospitals. *Science* **173**, 836 (1971).

25 Nikaido, H. & Vaara, M. Molecular basis of bacterial outer membrane permeability. *Microbiology and Molecular Biology Reviews* **49**, 1 (1985).

26 Chaplin, C. Bacterial resistance to quaternary ammonium disinfectants. *Journal of bacteriology* **63**, 453 (1952).

27 Mayor, S. News: Hospital acquired infections kill 5000 patients a year in England. *Bmj* **321**, 1370 (2000).

28   Jampala, S. N., Sarmadi, M., Somers, E., Wong, A. & Denes, F. Plasma-Enhanced Synthesis of Bactericidal Quaternary Ammonium Thin Layers on Stainless Steel and Cellulose Surfaces. *Langmuir* **24**, 8583-8591 (2008).

29   Verran, J., Airey, P., Packer, A. & Whitehead, K. A. Microbial retention on open food contact surfaces and implications for food contamination. *Advances in applied microbiology* **64**, 223-246 (2008).

30   Von Eiff, C., Kohnen, W., Becker, K. & Jansen, B. Modern strategies in the prevention of implant-associated infections. *International journal of artificial organs* **28**, 1146-1156 (2005).

31   Raad, I., Hachem, R., Zermeno, A., Dumo, M. & Bodey, G. P. In vitro antimicrobial efficacy of silver iontophoretic catheter. *Biomaterials* **17**, 1055-1059 (1996).

32   Palmer, J., Flint, S. & Brooks, J. Bacterial cell attachment, the beginning of a biofilm. *Journal of Industrial Microbiology and Biotechnology* **34**, 577-588 (2007).

33   Costerton, J. W., Stewart, P. S. & Greenberg, E. P. Bacterial biofilms: a common cause of persistent infections. *Science* **284**, 1318 (1999).

34   Costerton, J. *et al.* Biofilms, the customized microniche. *Journal of bacteriology* **176**, 2137 (1994).

35   Drenkard, E. Antimicrobial resistance of Pseudomonas aeruginosa biofilms. *Microbes and infection* **5**, 1213-1219 (2003).

36   Davies, D. G. *et al.* The involvement of cell-to-cell signals in the development of a bacterial biofilm. *Science* **280**, 295 (1998).

37   Jenkins, A. T. A., Buckling, A. & McGhee, M. Surface plasmon resonance shows that type IV pili are important in surface attachment by Pseudomonas aeruginosa. *Journal of the Royal Society Interface* **2**, 255-259 (2005).

38   Ramsey, M. M. & Whiteley, M. Pseudomonas aeruginosa attachment and biofilm development in dynamic environments. *Molecular microbiology* **53**, 1075-1087 (2004).

39   Cunliffe, D., Smart, C., Alexander, C. & Vulfson, E. Bacterial adhesion at synthetic surfaces. *Applied and environmental microbiology* **65**, 4995 (1999).

40   Katsikogianni, M. & Missirlis, Y. Concise review of mechanisms of bacterial adhesion to biomaterials and of techniques used in estimating bacteria-material interactions. *Eur Cell Mater* **8**, 37-57 (2004).

41   An, Y. H. & Friedman, R. J. Concise review of mechanisms of bacterial adhesion to biomaterial surfaces. *Journal of biomedical materials research* **43**, 338-348 (1998).

42   Persson, T., Givskov, M. & Nielsen, J. Quorum sensing inhibition: targeting chemical communication in Gramnegative bacteria. *Current medicinal chemistry* **12**, 3103-3115 (2005).

43   Chhabra, S. *et al.* Extracellular communication in bacteria. *The Chemistry of Pheromones and Other Semiochemicals II*, 117-128 (2005).

44   Doyle, R. J. *Microbial Growth in Biofilms: Special environments and physicochemical aspects*. Vol. 337 (Academic Pr, 2001).

45   Luppens, S. B. I., Reij, M. W., van der Heijden, R. W. L., Rombouts, F. M. & Abee, T. Development of a standard test to assess the resistance of Staphylococcus aureus biofilm cells to disinfectants. *Applied and environmental microbiology* **68**, 4194 (2002).

46   Wach, J. Y., Bonazzi, S. & Gademann, K. Antimicrobial surfaces through natural product hybrids. *Angewandte Chemie International Edition* **47**, 7123-7126 (2008).

47   Gabriel, M., Nazmi, K., Veerman, E. C., Amerongen, A. V. N. & Zentner, A. Preparation of LL-37-grafted titanium surfaces with bactericidal activity. *Bioconjugate Chem* **17**, 548-550 (2006).

48 Senuma, M., Tashiro, T., Iwakura, M., Kaeriyama, K. & Shimura, Y. Synthesis and antibacterial activity of copolymers having a quaternary ammonium salt side group. *Journal of Applied Polymer Science* **37** (1989).

49 Curtin, J. & Donlan, R. Using bacteriophages to reduce formation of catheter-associated biofilms by Staphylococcus epidermidis. *Antimicrobial agents and chemotherapy* **50**, 1268 (2006).

50 Capelli, C. C.   (Google Patents, 1990).

51 Krishnan, S., Weinman, C. J. & Ober, C. K. Advances in polymers for anti-biofouling surfaces. *Journal of Materials Chemistry* **18**, 3405-3413 (2008).

52 Nun, E. & Oles, M.   (Google Patents, 2002).

53 Shateri Khalil-Abad, M. & Yazdanshenas, M. E. Superhydrophobic antibacterial cotton textiles. *Journal of Colloid and Interface Science* **351**, 293-298 (2010).

54 Miyahara, Y., Mitamura, K., Saito, N. & Takai, O. Fabrication of microtemplates for the control of bacterial immobilization. *Journal of Vacuum Science & Technology A: Vacuum, Surfaces, and Films* **27**, 1183 (2009).

55 Höfer, D. Antimicrobial textiles–evaluation of their effectiveness and safety. *Biofunctional Textiles and the Skin. Burg G (ed). Curr Probl Dermatol. Basel, Karger* **33**, 42-50 (2006).

56 Ho, C. H., Tobis, J., Sprich, C., Thomann, R. & Tiller, J. C. Nanoseparated polymeric networks with multiple antimicrobial properties. *Advanced materials* **16**, 957-961 (2004).

57 Ioannou, C. J., Hanlon, G. W. & Denyer, S. P. Action of disinfectant quaternary ammonium compounds against Staphylococcus aureus. *Antimicrobial agents and chemotherapy* **51**, 296-306 (2007).

58 MacGregor, D. & Elliker, P. A comparison of some properties of strains of Pseudomonas aeruginosa sensitive and resistant to quaternary ammonium compounds. *Canadian journal of microbiology* **4**, 499-503 (1958).

59 Willis, A. E., Perham, R. N. & Wraith, D. Immunological properties of foreign peptides in multiple display on a filamentous bacteriophage. *Gene* **128**, 79-83 (1993).

60 Kim, S. W., Bae, Y. H. & Okano, T. Hydrogels: swelling, drug loading, and release. *Pharmaceutical research* **9**, 283-290 (1992).

61 Levy, I. et al. Chlorhexidine-impregnated dressing for prevention of colonization of central venous catheters in infants and children: a randomized controlled study. *The Pediatric infectious disease journal* **24**, 676 (2005).

62 Risbud, M. V. & Bhonde, R. R. Polyacrylamide-chitosan hydrogels: in vitro biocompatibility and sustained antibiotic release studies. *Drug Delivery* **7**, 69-75 (2000).

63 Man, G. et al. Potential wound dressing with improved antimicrobial property. *Journal of Applied Polymer Science* **105**, 1679-1686 (2007).

64 Darouiche, R. O. Anti-infective efficacy of silver-coated medical prostheses. *Clinical Infectious Diseases* **29**, 1371 (1999).

65 Yamamoto, K. & Ishihama, A. Transcriptional response of Escherichia coli to external zinc. *Journal of bacteriology* **187**, 6333 (2005).

66 Yamamoto, K. & Ishihama, A. Transcriptional response of Escherichia coli to external copper. *Molecular microbiology* **56**, 215-227 (2005).

67 Matsumura, Y., Yoshikata, K., Kunisaki, S. & Tsuchido, T. Mode of bactericidal action of silver zeolite and its comparison with that of silver nitrate. *Applied and environmental microbiology* **69**, 4278 (2003).

68 Gibbins, B. & Warner, L. The role of antimicrobial silver nanotechnology. *Medical Device & Diagnostic Industry Magazine*, 112 (2005).

69 Russell, A. Biocide use and antibiotic resistance: the relevance of laboratory findings to clinical and environmental situations. *The Lancet Infectious Diseases* **3**, 794-803 (2003).

70 Russell, A. & Hugo, W. 7 Antimicrobial Activity and Action of Silver. *Progress in medicinal chemistry* **31**, 351-370 (1994).

71 Feng, Q. et al. A mechanistic study of the antibacterial effect of silver ions on Escherichia coli and Staphylococcus aureus. *Journal of biomedical materials research* **52**, 662-668 (2000).

72 Schapiro, J. M., Libby, S. J. & Fang, F. C. Inhibition of bacterial DNA replication by zinc mobilization during nitrosative stress. *Proceedings of the National Academy of Sciences of the United States of America* **100**, 8496 (2003).

73 Sheng, J., Nguyen, P. & Marquis, R. E. Multi-target antimicrobial actions of zinc against oral anaerobes. *Archives of Oral Biology* **50**, 747-757 (2005).

74 Phan, T. N., Buckner, T., Sheng, J., Baldeck, J. & Marquis, R. Physiologic actions of zinc related to inhibition of acid and alkali production by oral streptococci in suspensions and biofilms. *Oral microbiology and immunology* **19**, 31-38 (2004).

75 Osorio, R. M., Hefti, A., Vertucci, F. J. & Shawley, A. L. Cytotoxicity of endodontic materials. *Journal of Endodontics* **24**, 91-96 (1998).

76 Rustogi, R., Mill, J., Fraser, J. & Kimble, R. The use of Acticoat (TM) in neonatal burns. *Burns* **31**, 878-882 (2005).

77 Fosmire, G. J. Zinc toxicity. *The American journal of clinical nutrition* **51**, 225 (1990).

78 Falchuk, K. H. The molecular basis for the role of zinc in developmental biology. *Molecular and cellular biochemistry* **188**, 41-48 (1998).

79 Baldwin, S. *et al.* Skin benefits from continuous topical administration of a zinc oxide/petrolatum formulation by a novel disposable diaper. *Journal of the European Academy of Dermatology and Venereology* **15**, 5-11 (2001).

80 Harrison, J. J. *et al.* Copper and quaternary ammonium cations exert synergistic bactericidal and antibiofilm activity against Pseudomonas aeruginosa. *Antimicrobial agents and chemotherapy* **52**, 2870-2881 (2008).

81 Sant, S., Gill, K. & Burrell, R. Nanostructure, dissolution and morphology characteristics of microcidal silver films deposited by magnetron sputtering. *Acta Biomaterialia* **3**, 341-350 (2007).

82 Favia, P. & D'AGOSTINO, R. Plasma processed surfaces for biomedical devices: PEO-like, Ag/PEO-like,-COOH functional and micro-patterned coatings: Procédés plasmas froids. *Le Vide* **57**, 40-48 (2002).

83 Wang, J. *et al.* The biomedical properties of polyethylene terephthalate surface modified by silver ion implantation. *Nuclear Instruments and Methods in Physics Research Section B: Beam Interactions with Materials and Atoms* **257**, 141-145 (2007).

84 Sarangapani, S.   (Google Patents, 2008).

85 Balogh, L., Swanson, D. R., Tomalia, D. A., Hagnauer, G. L. & McManus, A. T. Dendrimer-silver complexes and nanocomposites as antimicrobial agents. *Nano letters* **1**, 18-21 (2001).

86 Furno, F. *et al.* Silver nanoparticles and polymeric medical devices: a new approach to prevention of infection? *Journal of Antimicrobial Chemotherapy* **54**, 1019 (2004).

87 Vasilev, K. *et al.* Antibacterial surfaces by adsorptive binding of polyvinyl-sulphonate-stabilized silver nanoparticles. *Nanotechnology* **21**, 215102 (2010).

88 Jenkins, A. *et al.* Pulsed plasma deposited maleic anhydride thin films as supports for lipid bilayers. *Langmuir* **16**, 6381-6384 (2000).

89 Vasilev, K., Britcher, L., Casanal, A. & Griesser, H. J. Solvent-induced porosity in ultrathin amine plasma polymer coatings. *The Journal of Physical Chemistry B* **112**, 10915-10921 (2008).

90 Chapman, C. L., Bhattacharyya, D., Eberhart, R. C., Timmons, R. B. & Chuong, C. J. Plasma polymer thin film depositions to regulate gas permeability through nanoporous track etched membranes. *Journal of Membrane Science* **318**, 137-144 (2008).

91 Sakata, J., Yamamoto, M. & Hirai, M. Plasma polymerized membranes and gas permeability. II. *Journal of Applied Polymer Science* **31**, 1999-2006 (1986).

92 Yamamoto, M., Sakata, J. & Hirai, M. Plasma polymerized membranes and gas permeability. I. *Journal of Applied Polymer Science* **29**, 2981-2987 (1984).

93 Sakata, J., Hirai, M. & Yamamoto, M. Plasma-polymerized membranes and gas permeability III. *Journal of Applied Polymer Science* **34**, 2701-2711 (1987).

94 Ogawa, M., Natsume, Y., Hirayama, T. & Sakata, H. Preparation and electrical properties of undoped zinc oxide films by CVD. *Journal of Materials Science Letters* **9**, 1351-1353 (1990).

95 Duque, L. & Förch, R. Plasma Polymerization of Zinc Acetyl Acetonate for the Development of a Polymer-based Zinc Release System. *Plasma Processes and Polymers* (2011).

96 Hamerli, P., Weigel, T., Groth, T. & Paul, D. Surface properties of and cell adhesion onto allylamine-plasma-coated polyethylenterephtalat membranes. *Biomaterials* **24**, 3989-3999 (2003).

97 Yang, Z. et al. The covalent immobilization of heparin to pulsed-plasma polymeric allylamine films on 316L stainless steel and the resulting effects on hemocompatibility. *Biomaterials* **31**, 2072-2083 (2010).

98 Schonherr, H. *et al.* Distributions of functional groups in plasma polymerized allylamine films by scanning force microscopy using functionalized probe tips. *Chem. Mater* **12**, 3689-3694 (2000).

99 Förch, R. *et al.* Recent and expected roles of plasma-polymerized films for biomedical applications. *Chemical Vapor Deposition* **13**, 280-294 (2007).

100 Robson, M., Stenberg, B. & Heggers, J. Wound healing alterations caused by infection. *Clinics in plastic surgery* **17**, 485 (1990).

101 Robson, M. C. WOUND INFECTION:: A Failure of Wound Healing Caused by an Imbalance of Bacteria. *Surgical Clinics of North America* **77**, 637-650 (1997).

102 Percival, S. L., Bowler, P. & Woods, E. J. Assessing the effect of an antimicrobial wound dressing on biofilms. *Wound repair and regeneration* **16**, 52-57 (2008).

103 Geiger, B., Bershadsky, A., Pankov, R. & Yamada, K. M. Extracellular matrix–cytoskeleton crosstalk. *Nat Rev Mol Cell Biol* **2**, 793-805 (2001).

104 Juliano, R. & Haskill, S. Signal transduction from the extracellular matrix. *The Journal of cell biology* **120**, 577-585 (1993).

105 Grinnell, F., Billingham, R. E. & Burgess, L. Distribution of fibronectin during wound healing in vivo. *Journal of Investigative Dermatology* **76**, 181-189 (1981).

106 Alovskaya, A., Alekseeva, T., Phillips, J., King, V. & Brown, R. Fibronectin, Collagen, Fibrin-Components of Extracellular Matrix for Nerve Regeneration. *Top Tissue Eng. Milton Keynes: Biomaterials and Tissue Eng Group*, 1-26 (2007).

107 Leahy, D. J., Aukhil, I. & Erickson, H. P. 2.0 Å crystal structure of a four-domain segment of human fibronectin encompassing the RGD loop and synergy region. *Cell* **84**, 155-164 (1996).

108 Zelzer, M., Albutt, D., Alexander, M. R. & Russell, N. A. The Role of Albumin and Fibronectin in the Adhesion of Fibroblasts to Plasma Polymer Surfaces. *Plasma Processes and Polymers*, n/a-n/a, doi:10.1002/ppap.201100054 (2011).

109 Yoshinari, M. *et al.* Immobilization of Fibronectin onto Organic Hexamethyldisiloxane Coatings with Plasma Surface Modification. *Journal of Oral Tissue Engineering* **1**, 69-79 (2004).

110 Seeger, J. & Klingman, N. Improved endothelial cell seeding with cultured cells and fibronectin-coated grafts. *J Surg Res* **38**, 641-647 (1985).

111 Rocco, M., Infusini, E., Daga, M., Gogioso, L. & Cuniberti, C. Models of fibronectin. *EMBO J* **6**, 2343-2349 (1987).

112 Pierschbacher, M. D. & Ruoslahti, E. Cell attachment activity of fibronectin can be duplicated by small synthetic fragments of the molecule. *Nature* **309**, 30 (1984).

113 Lotz, A., Heller, M., Brieger, J., Gabriel, M. & Förch, R. Derivatization of Plasma Polymerized Thin Films and Attachment of Biomolecules to Influence HUVEC-Cell Adhesion. *Plasma Processes and Polymers* **9**, 10-16 (2012).

114 Gigout, A., Ruiz, J. C., Wertheimer, M. R., Jolicoeur, M. & Lerouge, S. Nitrogen Rich Plasma Polymerized Coatings on PET and PTFE Surfaces Improve Endothelial Cell Attachment and Resistance to Shear Flow. *Macromolecular Bioscience* **11**, 1110-1119 (2011).

115 Gabriel, M., Amerongen, G., Van Hinsbergh, V., Amerongen, A. & Zentner, A. Direct grafting of RGD-motif-containing peptide on the surface of polycaprolactone films. *Journal of Biomaterials Science - Polymer Edition* **17**, 567-577 (2006).

116 Langmuir, I. Oscillations in ionized gases. *Proceedings of the National Academy of Sciences of the United States of America* **14**, 627 (1928).

117 d'Angelo, N. Low-frequency electrostatic waves in dusty plasmas. *Planetary and Space Science* **38**, 1143-1146 (1990).

118 Yasuda, H. *Plasma Polymerization*. (Academic Press Inc., 1985).

119 Yasuda, H. & Hsu, T. Some aspects of plasma polymerization investigated by pulsed RF discharge. *Journal of polymer science: Polymer chemistry edition* **15**, 81-97 (1977).

120 Koidl, P., Wild, C., Dischler, B., Wagner, J. & Ramsteiner, M. 41-70 (Trans Tech Publ).

121 Liston, E., Martinu, L. & Wertheimer, M. Plasma surface modification of polymers for improved adhesion: a critical review. *Journal of Adhesion Science and Technology* **7**, 1091-1127 (1993).

122 Poncin-Epaillard, F. & Legeay, G. Surface engineering of biomaterials with plasma techniques. *Journal of Biomaterials Science, Polymer Edition* **14**, 1005-1028 (2003).

123 Mahan, J. E. Physical vapor deposition of thin films. *Physical Vapor Deposition of Thin Films, by John E. Mahan, pp. 336. ISBN 0-471-33001-9. Wiley-VCH, January 2000.* **1** (2000).

124 Arima, Y. & Iwata, H. Effect of wettability and surface functional groups on protein adsorption and cell adhesion using well-defined mixed self-assembled monolayers. *Biomaterials* **28**, 3074-3082 (2007).

125 Yasuda, T., Okuno, T. & Yasuda, H. Contact angle of water on polymer surfaces. *Langmuir* **10**, 2435-2439 (1994).

126 Wang, J. H., Claesson, P., Parker, J. & Yasuda, H. Dynamic contact angles and contact angle hysteresis of plasma polymers. *Langmuir* **10**, 3887-3897 (1994).

127 Dussan, E. On the spreading of liquids on solid surfaces: static and dynamic contact lines. *Annual Review of Fluid Mechanics* **11**, 371-400 (1979).

128 Hesse, M., Meier, H. & Zeeh, B. *Spektroskopische Methoden in der organischen Chemie.* (Georg Thieme Verlag, 2005).

129 Ekgasit, S., Thammacharoen, C. & Knoll, W. Surface plasmon resonance spectroscopy based on evanescent field treatment. *Anal. Chem* **76**, 561-568 (2004).

130 Homola, J., Yee, S. S. & Gauglitz, G. Surface plasmon resonance sensors: review. *Sensors & Actuators: B. Chemical* **54**, 3-15 (1999).

131 Fassel, V. A. & Kniseley, R. N. Inductively coupled plasma. Optical emission spectroscopy. *Analytical Chemistry* **46**, 1110-1120 (1974).

132 Scott, R. H., Fassel, V. A., Kniseley, R. N. & Nixon, D. E. Inductively coupled plasma-optical emission analytical spectrometry. *Analytical Chemistry* **46**, 75-80 (1974).

133 Butler, J., Joy, D., Bradley, G. & Krause, S. Low-voltage scanning electron microscopy of polymers. *Polymer* **36**, 1781-1790 (1995).

134 Beamson, G. & Briggs, D. *High resolution XPS of organic polymers: the Scienta ESCA300 database.* (Wiley Chichester, 1992).

135 Beamson, G. *et al.* Performance and application of the scienta ESCA300 spectrometer. *Surface and Interface Analysis* **15** (1990).

136 Forch, R., Schönherr, H. & Jenkins, A. T. A. *Surface design: applications in bioscience and nanotechnology.* Vol. 471 (Wiley-VCH, 2009).

137 Briggs, D. *Surface analysis of polymers by XPS and static SIMS.* (Cambridge University Press, 1998).

138 Bryson, C. E. Surface potential control in XPS. *Surface Science* **189**, 50-58 (1987).

139 Barth, G., Linder, R. & Bryson, C. Advances in charge neutralization for XPS measurements of nonconducting materials. *Surface and Interface Analysis* **11** (1988).

140 Favia, P., Stendardo, M. V. & d'Agostino, R. Selective grafting of amine groups on polyethylene by means of NH 3− H 2 RF glow discharges. *Plasmas and Polymers* **1**, 91-112 (1996).

141 Ruiz, J. C., St Georges Robillard, A., Thérésy, C., Lerouge, S. & Wertheimer, M. R. Fabrication and Characterisation of Amine Rich Organic Thin Films: Focus on Stability. *Plasma Processes and Polymers* (2010).

142 Gabriel, M., Dahm, M. & Vahl, C. F. Wet-chemical approach for the cell-adhesive modification of polytetrafluoroethylene. *Biomedical Materials* **6**, 035007 (2011).

143 Philip, N. *et al.* The respective roles of UV photons and oxygen atoms in plasma sterilization at reduced gas pressure: the case of N 2-O 2 mixtures. *IEEE Transactions on Plasma Science* **30**, 1429-1436 (2002).

144 Gallino, E., Massey, S., Tatoulian, M. & Mantovani, D. Plasma polymerized allylamine films deposited on 316L stainless steel for cardiovascular stent coatings. *Surface and Coatings Technology* (2010).

145 Harsch, A., Calderon, J., Timmons, R. & Gross, G. Pulsed plasma deposition of allylamine on polysiloxane: a stable surface for neuronal cell adhesion. *Journal of Neuroscience Methods* **98**, 135-144 (2000).

146 Rebl, H. *et al.* Positively Charged Material Surfaces Generated by Plasma Polymerized Allylamine Enhance Vinculin Mobility in Vital Human Osteoblastss. *Advanced Engineering Materials* (2010).

147 Dyer, M., Ainslie, K. & Pishko, M. Protein adhesion on silicon-supported hyperbranched poly (ethylene glycol) and poly (allylamine) thin films. *Mater. Res* **23**, 1384-1393 (1989).

148 Basarir, F., Cuong, N., Song, W. & Yoon, T. 61-66 (Wiley Online Library).

149 Finke, B. *et al.* Plasma Polymer Coating of Titanium for Improved Bone Implants. *NBC-14th Nordic-Baltic Conference on Biomedical Engineering and Medical Physics 16-20 June 2008; Riga, Latvia*, 30 (2008).

150 Li, L. *et al.* Combined continuous wave and pulsed plasma modes: For more stable interfaces with higher functionality on metal and semiconductor surfaces. *Plasma Processes and Polymers* **6**, 615-619 (2009).

151 Chifen, A. N. *Plasma Polymerized Biofunctional Multilayers Based on SiO2-like Adhesion Promoting Film*, (2007).

152 Chifen, A. N. *et al.* Attachment and Phospholipase A 2-Induced Lysis of Phospholipid Bilayer Vesicles to Plasma-Polymerized Maleic Anhydride/SiO 2 Multilayers. *Langmuir* **23**, 6294-6298 (2007).

153 Pihan, S. *Silizium Wafer Bonding mittels Si-haltiger PECVD-Beschichtungen*, FH Wiesbaden, (2008).

154 Hayakawa, T., Yoshinari, M. & Nemoto, K. Characterization and protein-adsorption behavior of deposited organic thin film onto titanium by plasma polymerization with hexamethyldisiloxane. *Biomaterials* **25**, 119-127 (2004).

155 Wei, J. *et al.* Adhesion of mouse fibroblasts on hexamethyldisiloxane surfaces with wide range of wettability. *Journal of Biomedical Materials Research Part B: Applied Biomaterials* (2007).

156 Krishnamurthy, V., Kamel, I. & Wei, Y. Analysis of plasma polymerization of allylamine by FTIR. *Journal of Polymer Science. Part A. Polymer Chemistry* **27**, 1211-1224 (1989).

157 Yang, Z. *et al.* Pulsed-Plasma Polymeric Allylamine Thin Films. *Plasma Processes and Polymers* **6**, 498-505 (2009).

158 Choukourov, A. *et al.* Mechanistic studies of plasma polymerization of allylamine. *Journal of Physical Chemistry B* **109**, 23086-23095 (2005).

159 Chu, L. Q., Knoll, W. & Forch, R. Stabilization of plasma-polymerized allylamine films by ethanol extraction. *Langmuir* **22**, 5548-5551 (2006).

160 Calderon, J. G., Harsch, A., Gross, G. W. & Timmons, R. B. Stability of plasma-polymerized allylamine films with sterilization by autoclaving. *Journal of biomedical materials research* **42**, 597-603 (1998).

161 Vasilev, K., Michelmore, A., Griesser, H. J. & Short, R. D. Substrate influence on the initial growth phase of plasma-deposited polymer films. *Chem. Commun.*, 3600-3602 (2009).

162 Baquey, C., Palumbo, F., Porte-Durrieu, M.C., Legeay, G., Tressaud, A., d'Agostino, R. Plasma treatment of expanded PTFE offers a way to a biofunctionalization of its surface. *Nuclear Instruments and Methods in Physics Research B* **151**, 255-262 (1999).

163 Nelea, V., Nakano, Y. & Kaartinen, M. T. Size distribution and molecular associations of plasma fibronectin and fibronectin crosslinked by transglutaminase 2. *The Protein Journal* **27**, 223-233 (2008).

164 Hersel, U., Dahmen, C. & Kessler, H. RGD modified polymers: biomaterials for stimulated cell adhesion and beyond. *Biomaterials* **24**, 4385-4415 (2003).

165 McPherson, T. B. & Badylak, S. F. Characterization of fibronectin derived from porcine small intestinal submucosa. *Tissue Engineering* **4**, 75-83 (1998).

166 Tarducci, C., Kinmond, E., Badyal, J., Brewer, S. & Willis, C. Epoxide-functionalized solid surfaces. *Chemistry of materials* **12**, 1884-1889 (2000).

167 Evtushenko, Y., Ivanov, V. & Zaitsev, B. Determination of epoxide and hydroxyl groups in epoxide resins by IR spectrometry. *Journal of Analytical Chemistry* **58**, 347-350 (2003).

168 Salacinski, H. J., Tiwari, A., Hamilton, G. & Seifalian, A. M. Cellular engineering of vascular bypass grafts: Role of chemical coatings for enhancing endothelial cell attachment. *Medical and Biological Engineering and Computing* **39**, 609-618 (2001).

169 Tiwari, A., Salacinski, H., Hamilton, G. & Seifalian, A. Tissue Engineering of Vascular Bypass Grafts: Role of Endothelial Cell Extraction. *European Journal of Vascular & Endovascular Surgery* **21**, 193-201 (2001).

170 Walluscheck, K., Steinhoff, G., Kelm, S. & Haverich, A. Improved endothelial cell attachment on ePTFE vascular grafts pretreated with synthetic RGD-containing peptides. *European Journal of Vascular & Endovascular Surgery* **12**, 321-330 (1996).

171 Roberts, M., Bentley, M. & Harris, J. Chemistry for peptide and protein PEGylation. *Advanced drug delivery reviews* **54**, 459-476 (2002).

172 Gabriel M, N. K., Dahm M, Zentner A, Vahl CF, Strand D. Covalent RGD Modification of the Inner Pore Surface of Polycaprolactone Scaffolds. *J Biomater Sci Polym Ed* **22** (2011).

173 Schuler, M. *et al.* Biomimetic modification of titanium dental implant model surfaces using the RGDSP-peptide sequence: A cell morphology study. *Biomaterials* **27**, 4003-4015 (2006).

174 Ovadia, O. *et al.* The effect of multiple N-methylation on intestinal permeability of cyclic hexapeptides. *Molecular Pharmaceutics* (2011).

175 Takada, Y., Ye, X. & Simon, S. The integrins. *Genome biology* **8**, 215 (2007).

176 Intranuovo, F. *et al.* Plasma modification of PCL porous scaffolds fabricated by Solvent Casting/Particulate Leaching for Tissue Engineering. *bone* **15**, 16.

177 Söderberg, T. A. *et al.* Antibacterial effect of zinc oxide in vitro. *Scandinavian Journal of Plastic and Reconstructive Surgery and Hand Surgery* **24**, 193-197 (1990).

178 Hassan, M. T. *et al.* Identification of a gene cluster, czr, involved in cadmium and zinc resistance in Pseudomonas aeruginosa. *Gene* **238**, 417-425 (1999).

179 Foerch, R., Duque, L. & Lotz, A. Multifunctional Coating With Antimicrobial Activity And Cell Adhesion Regulating Surface Characteristics As Well As A Method For Preparing The Same. (2012).

180 Inagaki, N., Tasaka, S. & Nozue, Y. Plasma polymerization of metal acetylacetonates and application for gas sensor devices. *Journal of Applied Polymer Science* **45**, 1041-1048 (1992).

181 Ip, M., Lui, S. L., Poon, V. K. M., Lung, I. & Burd, A. Antimicrobial activities of silver dressings: an in vitro comparison. *Journal of medical microbiology* **55**, 59-63 (2006).

182 Miura, N. & Shinohara, Y. Cytotoxic effect and apoptosis induction by silver nanoparticles in HeLa cells. *Biochemical and biophysical research communications* **390**, 733-737 (2009).

183 Liau, S., Read, D., Pugh, W., Furr, J. & Russell, A. Interaction of silver nitrate with readily identifiable groups: relationship to the antibacterialaction of silver ions. *Letters in Applied Microbiology* **25**, 279-283 (1997).

184 Bridges, K., Kidson, A., Lowbury, E. & Wilkins, M. Gentamicin-and silver-resistant pseudomonas in a burns unit. *British medical journal* **1**, 446-449 (1979).

i want morebooks!

Buy your books fast and straightforward online - at one of world's fastest growing online book stores! Environmentally sound due to Print-on-Demand technologies.

Buy your books online at
www.get-morebooks.com

Kaufen Sie Ihre Bücher schnell und unkompliziert online – auf einer der am schnellsten wachsenden Buchhandelsplattformen weltweit! Dank Print-On-Demand umwelt- und ressourcenschonend produziert.

Bücher schneller online kaufen
www.morebooks.de

VDM Verlagsservicegesellschaft mbH
Heinrich-Böcking-Str. 6-8   Telefon: +49 681 3720 174   info@vdm-vsg.de
D - 66121 Saarbrücken   Telefax: +49 681 3720 1749   www.vdm-vsg.de

Printed by Books on Demand GmbH, Norderstedt / Germany